THE UAPS MANUAL VOLUME 1

Plant Tissue Culture Procedures

Wyc College Press

CONTRIBUTORS

Kodi Kandasamy
Course Tutor of Short Training Course in Plant Tissue Culture and its uses in Micropropagation and Plant Biotechnology (1991-1995)

Sinclair Mantell
Reader in Horticulture (1983-1995)

Jennet Blake
Senior Research Fellow (1983-1994)

David Newton
Course Tutor (1989-1991)

Rosalind Harris
Course Tutor (1985-1989)

Mrs Susan Farris
UAPS Laboratory Technician (1988-1995)

FOREWORD

This first volume of The UAPS Manual, like its sister Volume 2, derives from the result of ten years of teaching the Wye College Short Course entitled "Plant tissue culture and its applications in micropropagation and plant biotechnology". It is the product of several upgraded versions of class exercises in plant cell and tissue culture techniques that have been used to guide trainees through their three months of attachment to the Unit for Advanced Propagation Systems. The Manual therefore contains reference to equipment and facilities in the UAPS laboratory. Nevertheless, it is hoped that the information it contains will be useful in other plant tissue culture laboratory situations worldwide since over seventy trainees have passed through the UAPS Short Course in plant tissue culture at Wye College and are now working in many different countries. It seems likely therefore that the exercises will prove valuable for beginners and to technical staff undergoing in- service training activities.

Finally, it is hoped that this Manual will be useful to both laboratory managers and plant tissue culture operators in their efforts to micropropagate and genetically manipulate crop plants to the benefit of mankind. Any suggestions for improvements will be gratefully received by the UAPS staff involved in the production of this second edition of The UAPS Manual.

GOLDEN RULES IN A PLANT TISSUE CULTURE LABORATORY

Legal requirement: All users (that includes all staff, students, trainees and visiting scientist ie EVERYONE!) are required by law and common sense to complete a COSHH (Control Of Substances Harzadous to Health) assessment form which can be obtained from UAPS academic and technical staff (a copy of definitions and background information on COSHH is presented in Appendix 1.

1. Keep all areas clean and tidy: especially Culture Transfer Rooms. Avoid bringing any living plant materials into laboratory areas without being first surface sterilised. Wear clean laboratory coats and dirt-free footwear at all times. No soil to be introduced into the laboratory areas.

2. Leave all items and equipment after use in a condition that you yourself would like to find them in. This applies especially to laminar flow cabinets, balances and bench work surfaces.

3. Washing-up must be done on the same day as it is created. Contaminated cultures should be autoclaved first before washing up is carried out in order to reduce air borne spore levels.

4. When items and chemicals are running low, place a request immediately for these on the 'orders' list.

5. Switch off all electrical apparatus after use (**except pH meter**).

6. Place a card/paper label on all trays indicating the plant species in culture, the treatment, the date of subculture and most importantly **YOUR NAME**.

7. Do not write with permanent ink pens on polycarbonate or metal caps. Use easy to remove labels. This applies also to glassware used for culture work.

8. Keep accurate photographic records of the progress of tissue cultures.

9. Be honest about contamination: please inform a member of technical staff if contamination is particularly severe: problems from mites or other living organisms (like thrips) capable of spreading the contamination between cultures and rooms could be the cause and these matters have to be dealt with expeditiously.

10. Please be as ECONOMICAL as possible with all consumables, especially with alcohol. Swab benches and other surfaces with cheaper alternatives to alcohol like hospital/domestic disinfectants such as "Pursue"

CONTENTS

Page

Forward

Section 1	**UAPS CODE OF PRACTICE**	1

Section 2   **BACKGROUND INFORMATION ON PLANT TISSUE CULTURE**

     2.1.   General techniques and basic principles   9

     2.2.   Micropropagation/Cloning/Production of disease-free plants/Germplasm storage   11

     2.3.   Application of plant tissue and cell culture to crop breeding/agriculture   13

     2.4.   Application of plant tissue culture to production of biochemicals   14

     2.5.   General plant biotechnology texts   15

Section 3   **PREPARATION**

     3.1.   Care of mother stock plants and preferable sources of explants   20

     3.2.   Setting up the flow cabinet   24

     3.3   Surface sterilisation of explants   25

Section 4   **PRACTICAL EXERCISES**

Exercise 1: Introduction to procedures

     4.1.1.   Medium preparation   28

     4.1.2.   Calculation exercise   32

     4.1.3.   Surface sterilisation exercise   33

     4.1.4   Dissection and culturing of explants   35

Exercise 2: Establishing callus and cell suspension cultures

4.2.1. Callus initiation	36
4.2.2 Cell suspension initiation	41

Exercise 3: Micropropagation: Stages I, II and III

4.3.1. Meristem and shoot apex cultures	43
4.3.2. Rooting exercise	47

Exercise 4: Micropropagation: Stage IV

4.4.1 Weaning of micropropagated plantlets / acclimatization	49

Exercise 5: Morphogenesis

4.5.1. Organogenesis (direct/indirect)	51
4.5.2. Somatic embryogenesis	53

Exercise 6: Haploid plant production

4.6.1. Anther and pollen culture	55

Exercise 7: *In vitro* seed culture

4.7.1. Seed germination	67
4.7.2. Embryo rescue	68

Exercise 8: Microdissection

4.8.1. Micrografting	71

Exercise 9: Protoplast culture techniques

4.9.1.	Protoplast isolation and culture	74
4.9.2.	Growth and development assessment	83
4.9.3.	Protoplast fusion	85

Exercise 10: Problems in micropropagation

4.10.1.	Vitrification	89
4.10.2.	Browning	94

Exercise 11: Immobilisation

4.11.1.	Encapsulation of somatic embryos	97
4.11.2.	Immobilisation of embryoids or cells in agarose	101

Section 5	**GENERAL INFORMATION**	
5.1.	Plant growth substances and plant growth responses	103
5.2.	Role of vitamins in plants	105
5.3.	Quantification of *in vitro* plant production at stages II, III & IV	106
5.4.	Statistics	114
5.5.	Laboratory layout, equipment and costing	117
5.6.	Laboratory equipment suppliers	123

FIGURES

Figure 1. Development of micropropagation (1838-1994) 22

Figure 2. Micropropagation cycle 23

Figure 3. Haploid plant production in rapeseed by microspore culture 65

Figure 4. Graph showing the growth / multiplication rate of micropropagated plants 111

Figure 5. Floor plan of a typical plant tissue culture laboratory 118

TABLES

Table 1. Preparation of MS medium using stock solution — 31

Table 2. Inter conversion between mass/volume and molarity measurements — 32

Table 3. Composition of media for culture of anthers — 59

Table 4. The composition of NLN media — 62

Table 5. Protoplast isolation medium (TPIM) — 79

Table 6. Protoplast wash and culture medium (YPWCM) — 80

Table 7. Protoplast purification medium (TPPM) — 81

Table 8. PEG protoplast fusion medium (TPFM) — 88

Section 1: UAPS CODE OF PRACTICE

It is essential for the smooth running of the laboratory that the following Code of Practice be observed at all times by all those working in UAPS.

COSHH Regulations

Fill in a COSHH Form before starting any work in UAPS. A form can be obtained from any member of UAPS staff. Read the Safety Handbook attached to the UAPS Noticeboard at the entrance to the laboratories.

Decontamination Techniques and Aseptic Procedures

All successful tissue culture work is based upon the maintenance of sterile conditions at all stages of the laboratory procedures. The following things ensure that the tissue culture environment is as clean and well-ordered as possible under the prevailing conditions.

Decontamination of plant material

* Confine all fresh plant material to the dirty laboratory (Prep. Room) and never bring exposed living plant materials and soil into any of the laboratory areas. Dispose of unwanted plant parts immediately using the black bags mounted in dustbins.

* Wash your hands after working with fresh plant material to reduce the numbers of mites and other contaminating organisms entering the lab area.

* Use fresh bleach concentrates when making up chlorine solutions for plant sterilisation.

* Return bleach concentrates to the cold store immediately after use.

* Use sterilised distilled water for preparing bleach solutions.

Flow cabinet procedures

* Use fresh, recently sterilised (please date the bottles of water) distilled water (SDW) for making up 70% ethanol solutions. Refill and autoclave partly used bottles of SDW.

* Pre-sterilise instrument-dip bottles used in laminar flow cabinets so as to prevent introduction of contaminants from sterile washing procedures and to prevent cross contamination from any contaminated cultures during subculturing.

* Remember to wash your hands before commencing any *in vitro* work.

* Wear a clean laboratory coat and clean footware which has never been used outdoors or outside the UAPS tissue culture laboratory area.

* Sterilise all dissection instruments in a bead or clay steriliser for at least 20- 30 seconds before use. Preferably do not use alcohol(meths) burners for primary sterilisation since their flame is not sufficiently powerful for killing off heat resistant bacterial and fungal spores and in addition they pose a fire hazard.

* Do not cool dissection instruments by dipping in alcohol. Always air cool instruments on the toothed support stands provided. Note that instruments dipped in 70% EtOH alone are NOT - repeat NOT sterile since some microorganisms are known to be able to survive in alcohol solutions.

* Turn off flow cabinet fan before spraying 70% EtOH onto the bench. This avoids inadvertant inhalation of solvent.

* Use sterile paper towels for swabbing down all working surfaces. Tissues used straight from the roll are not automatically sterile.

* Ensure that autoclaved polypropylene disks and squares are dry before use by placing them into the dry heat oven. Heat resistant spores may germinate under damp conditions.

* Place wet and dry rubbish into separate plastic and paper bags, respectively, and dispose appropriately in the dustbins provided in either the dirty laboratory or the washing up area.

* Replace lids on used tubes, bottles etc. and wash immediately after use to prevent microbial contamination.

Treatment of contaminated cultures and instruments

* Wrap trays of contaminated cultures in greaseproof paper before autoclaving.

* All contaminated cultures, whether in plastic petri dishes, glass tubes or bottles, must be autoclaved for at least 40 minutes to kill any potential human pathogens contaminating cultures.

* Autoclave all instruments which have been in contact with contaminated substances e.g. hepa-filters, pipettes, sieves etc. It is essential to wash syringes and glass beakers which have been used for agar with HOT water as soon as you have finished dispensing media.

The washing procedure used for culture glassware in UAPS:

1. Fill the left hand sink with hot water, add 3 measures of cleaning fluid and wash glassware etc. thoroughly with a scouring pad.

2. Fill righthand sink with water and rinse glassware three times.

3. Then rinse glassware three times in tap distilled water (TDW).

4. Fill a plastic bowl with double distilled water (DDW) and rinse glassware once.

5. Place glassware in the ovens to dry; pipettes, volumetric flasks, metals and plastics should be placed in trays on the top of the oven to dry.

6. Keep the culture transfer rooms tidy at all times. The doors to each transfer room should be kept shut whenever possible. You will be allocated a cupboard or drawer space for your autoclaved medium. Prepared sterilised culture medium should be used within two weeks. If you should need to keep it for longer, regularly check that it has not become contaminated but in no circumstances keep for more than one month. Please dispose of any unwanted, unused medium and wash up the tubes.

7. All cultures placed in the growth room should be clearly labelled on a piece of coloured card showing 1) your name 2) date 3) plant species and 4) any experimental details you might wish to record. Label each culture tubes/vessels to avoid misplacing of individual cultures in an experiment. In the case of stock cultures labelling of each

tray might only be needed. Do not leave old cultures in the growth room. Cultures should be checked regularly for growth and development responses and to check for any incursions of mites. Only keep the cultures you wish to use and throw away and wash up any unwanted cultures. Should you need to keep old cultures for future work, you should check them regularly for signs of contamination. Infected cultures should be wrapped up and autoclaved before disposing.

> **NB: MITE infestations are a major source of serious contamination in plant tissue cultures and that these tend to be more common in old cultures: mites are attracted to decaying vegetative material.**

Use of chemicals

Take care when weighing out chemicals and making up stock solutions. Do not spill chemicals or solutions on balances or benches. Wipe up any spillage with water and/or 70% methylated spirit. Be particularly careful when using **hazardous** chemicals such as $HgCl_2$, 2,4-D and concentrated acids. Read the label on the bottle before weighing out a chemical substance to note any special statements such as 'do not breathe in dust', 'avoid contact with skin', and respond sensibly to these cautionary instructions and take the necessary precautions. Remember it is not only you who will be exposed to the hazard. Wear gloves and a mask where appropriate. Handle volatile substances in a fume cupboard. Make sure that those chemicals which need to be stored at 4°C are kept in the fridge/cold room and those requiring storage at 0°C are kept in the freezer. Check your stock solutions regularly. Discard any old, unwanted stock solutions. Any infected solution must be autoclaved and disposed of in an appropriate fashion. Toxic substances may have to be disposed of using special measures and these will be descibed to you at the time of filling in your COSSH forms.

If a chemical supply is running low, write the details of the chemical and if possible its catologue number and supplier up on the order form placed on the noticeboard next to the telephone in the preparation laboratory. This will ensure that the chemical supply is replaced in good time before current stocks are exhausted.

Use of glassware

The glass beakers, measuring cylinders, pipettes and syringes are expensive. Please be careful to wash individual items separately and avoid any breakages. If glassware does get broken it must be thrown away in the bin under the main sink in the broken glass box. Do not throw broken glass into any other bin as this will be a potential hazard to cleaning staff.

Use of equipment

If you have problems and you are not sure of how a piece of equipment is used, ask a member of staff before proceeding. Don't force any equipment. Do not attempt to use the following pieces of equipment until you have been shown by a member of staff:

*autoclaves - both the bench top and stand up ones

*pH meters - do not adjust the controls. These instruments are checked and

*calibrated regularly by UAPS technical staff

*OMT - 2 fluorescence / inverted microscope

*cameras

*fine balances (eg the Oertling R20 pan balance)

Note: Please report any breakages or malfunctions of equipment to a member of technical staff as soon as possible.

Dissection Instruments

Everyone is allocated a box containing dissection instruments. These should be looked after carefully. A deposit of £15 will be demanded upon issue of each set of instruments to cover the replacement of lost or damaged items. Instruments can be resterilised after each day's work in the ovens located in each of the culture transfer rooms. Each of the ovens is fitted with a timer which switches them on at midnight and off after 1½h running at 160°C.

Covers

Please make up your own polypropylene covers to replace those that you use in the cabinets.

Booking for space

This must be done before starting to work in the media preparation area or transfer rooms by filling the relevant forms located by the telephone and on each flow cabinet, respectively.

Unit Seminars and Meetings

During term time, seminars and general meetings are held at regular intervals in the Unit. These are for the benefit of everyone working in UAPS. It is strongly advised that everyone attend such meetings.

Working hours

The Russell Laboratories are open 07.00 - 17.00 weekdays Monday to Friday. Restricted access operation in UAPS requires use of a key for the two corridor doors. This can be obtained from the UAPS technical staff on payment of a small deposit. If you wish to use UAPS facilities outside normal working hours, it is necessary to obtain a front door combination from the chief technician or from the Head of Department's secretary. If you are working alone in the laboratory after hours take extra care, and do not attempt to work with hazardous chemicals or equipment with which you are unfamiliar. Keep hazardous work for regular working hours.

Library

If you borrow any books from the UAPS library, write your name down on the card inside the book and leave the card on the bookshelf. Books can only be borrowed overnight or over a weekend. A series of papers has also been compiled in Box files covering a range of plant tissue culture topics. If you wish to read them, please do so in the extension and do not remove any papers from the Unit. The papers are filed alphabetically by topic. If you do remove a paper, please make sure you return it to the correct place as this ensures that papers can be found easily for reference in the future.

Similarly if you remove cards from the index for photocopying, please replace them back correctly. An easy way to do this is to slip a blue card in the place of the card you remove, but please make sure to take out the blue card when you return the reference card to its correct position. If you are unsure about the position of a card, please place it in the draw labelled "Cards for filing".

Section 2: BACKGROUND INFORMATION ON PLANT TISSUE CULTURE

This is a list of titles (mainly post 1980) from which most of the available information on plant tissue culture and its applications in agriculture, horticulture and general biotechnology can be obtained. Updated each year, it should provide a useful source of reference. Note that most books are catalogued under either 581.0 (Plant Tissue Culture) 631 (Propagation) or 575 (Biotechnology).

At Wye College, the books and reference files are in more than one location. A prefix code is used to locate the materials as follows:

U/F   In UAPS Files

L/B   In book section of Library (on shelves) or in reserve in which case the books are held behind the main library desk (marked "*").

H   Horticulture Section

2.1. GENERAL TECHNIQUES AND BASIC PRINCIPLES

*L/B   Bengochea, T. & Dodds, J.J. (1986). **Plant protoplasts: a biotechnological tool for plant improvement.** London, Chapman and Hall.

*L/B   Bhojwani, S.S. & Razdan, M.K. (1983). **Plant tissue culture: theory and practice.** Amsterdam, Elsevier.

H   BIOTOL PROJECT (1991) **Biotechnological Innovations in Crop Improvement.** Butterworth-Heinemann. 290pp

L/B   Dixon, R.A. (Ed.) (1985). **Plant cell culture: a practical approach** Oxford, IRL Press.

*L/B   Dodds, J.H. & Roberts, L.W. (1985). **Experiments in Plant Tissue Culture.** Cambridge University Press. 2nd Edn.

*L/B   Evans, D.A. et al (1983). **Handbook of plant cell culture Vol. 1 Techniques for propagation and breeding.** New York, Macmillan.

*L/B   George, E.F. (1987). **Plant Culture Media: Vol. 1 Formulations and Uses.** Westbury, Exegetics Ltd.

L/B   Ingram, D.S. & Helgeson, J.P. (Eds.) (1980). **Tissue Culture Methods for Plant Pathologists.** Oxford, Blackwell Scientific.

L/B   Lindsey,K. (ed.) (1992) **Plant Tissue Culture Manual. Supplement 1.** Kluwer Academic Publishers, Dordrecht.

*L/B   Pierik, R.L.M. (1987). **In vitro culture of higher plants.** Dordrecht, Martinus-Nijhoff Publ.

L/B   Reinert, J. & Yeoman, M.M. (1982). **Plant Tissue Culture - a laboratory manual.** Berlin, Springer-Verlag.

L/B   Street, H.E. (Ed.) (1977). **Plant Tissue and Cell Culture.** Oxford, Blackwell Scientific. 2nd Edn.

*L/B   Vasil, I.K. (1984). **Cell culture and somatic cell genetics of plants. Vol. 1. Laboratory Procedures and their applications. Vol. 2. Cell growth, nutrition, cytodifferentiation.** London, Academic Press.

L/B   Yeoman, M.M. (1986). **Plant Cell Culture Technology.** Oxford, Blackwell Scientific Publications.

2.2. MICROPROPAGATION/CLONING/PRODUCTION OF DISEASE-FREE PLANTS/GERMPLASM STORAGE

*L/B   Abbot, A.J. & Atkin, R.K. (1987). **Improving vegetatively propagated crops.** London, Academic Press (631.53).

L/B   Acta Horticulturae, No. 212 (1987). **Symposium on *in vitro* problems related to mass propagation of horticultural plants. Vols. I, II** (631.53). Also A/H Nos. 226 and 227 (631.53).

L/B   Bajaj, Y.P.S. (1984). **Biotechnology in Agriculture and Forestry. Crops I, II, Trees I, II,** Berlin, Springer Verlag

L/B   Bhojwani, S.S. (ed) (1990). **Plant Tissue Culture: Applications and Limitations.** Amsterdam, Elsevier

L/B   Bonga, J.M. & Durzan, D.J. (eds) (1987) **Cell and Tissue Culture in Forestry Vols 1-3.** Martinus Nijhoff Publishers, Dordrecht.

L/B   Conger, B.C. (Ed.) (1981). **Cloning Agricultural Plants via *In Vitro* Techniques.** Boca Raton, Florida, CRC Press.

L/B   Debergh, P.C. & Zimmerman, R.H. (1991). **Micropropagation: technology and application.** Amsterdam, Kluwer.

U/F   George, E.F. & Sherrington, P.D. (1984). **Plant Propagation by Tissue Culture.** Basingstoke, Exegetics Ltd.

*L/B   Hartmann, H.T. & Kester, D.E. (1983). **Plant Propagation: principles and practices.** 4th Edn. Englewood Cliffs, Prentice Hall Inc. Chapter 16.

U/F   Hussey, G. (1983). *In vitro* propagation of agricultural and horticultural crops. In: Mantell, S.H. & Smith, H. (Eds.) **Plant Biotechnology,** SEB Seminar Ser. Vol. 18, pp. 111 - 138 Cambridge University Press.

L/B   Krikorian, A.D. (1982). Cloning higher plants from aseptically cultured tissues and cells. **Biol.Rev.57**, pp. 151 - 218.

U/F   Schafer-Menuhr (Ed.) (1985). **In vitro techniques Propagation and Long-term Storage.** Dordrecht, Martinus Nijhoff/Dr. W. Junk Publ.

H   Zakri, A.H., Normah, M.N. & Abdul Karim, A.G. (eds) (1991). **Conservation of Plant Genetic Resources through *in vitro* methods.** Forest Research Institute of Malaysia/ Malaysia National Committee on Plant Genetic Resources

*L/B   Zimmerman, R.H. et. al. (1986) (Eds.). **Tissue culture as a plant production system for horticultural crops.** Dordrecht, Martinus-Nijhoff/Dr. Junk.

2.3. APPLICATION OF PLANT TISSUE AND CELL CULTURE TO CROP BREEDING/AGRICULTURE

*L/B Ammirato, P.V., Sharp, W.R. et al (1983). **Handbook of plant cell culture. Vol. 2,3 & 4 Crop species.** New York, Macmillan.

L/B Bright, S.W.J. (1985). **Cereal tissue and cell culture.** Dordrecht, Martinus Nijhoff/Dr. W. Junk.

H Chapman, G.P., Mantell, S.H. & Daniels, R.W. (1985). **Experimental manipulation of ovule tissues.** London, Longmans.

L/B CIBA (1988). Applications of plant cell and tissue culture. New York, John Wiley & Sons (581.O).

L/B Day, P.R. (1986). **Biotechnology and crop improvement and protection.** Thornton Heath, BCPC. 575.14.

H Dodds, J.H. (Ed.) (1985). **Plant Genetic Engineering.** Cambridge, Cambridge University Press.

L/B Henke, R.R. (Ed.) (1985). **Tissue culture in Forestry and Agriculture.** New York, Plenum Press.

L/B International Atomic Energy Agency (1986). **Nuclear Techniques and in vitro culture for plant improvement.** Proc. Int. Symp. on Nuclear Techniques and *in vitro* culture. 19 - 23 Aug. 1985.

L/B Janick, J. (Ed.) **Plant Breeding Reviews**: Vols. 1,2,3 & 4 (continuing). Connecticut, AVI Publ. Co. Ltd.

L/B Mantell, S.H., Chapman, G.P. & Street, P.F.S. (1986). **The Chondriome - chloroplast and mitochondrial genomes**. London, Longmans.

H Puite, K.J., Dons, J.J.M., Huizing, H.J., Koornneef, M. & Krens, F.A. (eds) (1988). **Progress in Plant Protoplast Research.** Kluwer Academic Publishers. 414pp.

H Roca, W.M. & Mroginski, L.A. (1991). **Cultivo de Tejidos en la Agricultura: Fundamentos y Aplicaciones** CIAT Publication No 151, CIAT, Colombia 969 pp.

*L/B Semal, J. (1986) (Ed.). **Somaclonal variations and crop improvement**. Proc. of seminar in the CEC programme. Gembloux 3-5 Sept. 1985.

H Tomes, D.T., Ellis, B.E., Harney, P.M., Kasha, K.J. & Peterson, R.L. (Eds.) (1982). **Application of plant cell and tissue culture to Agriculture and Industry**. Guelph, University of Guelph.

*L/B Withers, L. & Alderson, P.G. (Eds.) (1986). **Plant tissue culture and its agricultural applications**. London, Butterworths.

2.4. APPLICATION OF PLANT TISSUE CULTURE TO PRODUCTION OF BIOCHEMICALS

*L/B Constabel, F. (1988). **Cell culture and somatic cell genetics of plants: Vol. 5. Phytochemicals in plant cell culture**, ed. F. Constabel & I. Vasil. London, Academic Press.

U/F Morris, P., Scragg, A.H., et al (1986). **Secondary metabolism in plant cell cultures**. Cambridge University Press.

2.5. GENERAL PLANT BIOTECHNOLOGY TEXTS

L/B   AFRC. (1987). **Biotechnology in agriculture**. London, AFRC (P.575.14:630).

*L/B   Farrington, J. (1989). **Agricultural Biotechnology: Prospects for the Third World**. ODI Publication. 88 pp.

L/B   Hedin, P.A., Menn, J.J. & Hollingworth, R.M. (1988). **Biotechnology for Crop Protection**. ACS Symposium Series No. 379 (575.14).

L/B   IRRI (International Rice Research Institute). (1985). **Biotechnology in international agricultural research**. Philippines, IRRI.

L/B   ICRISAT (1988). **Biotechnology in Tropical Crop Improvement** Edited by de Wet, J.M.J.

L/B   Jacobsson, S. (1986). **The Biotechnological Challenge**. Cambridge, Cambridge University Press.

L/B   Joffe, S. **Biotechnology and third world agriculture: a literature review and report**. Sussex, IDS. (Fol. 575.14).

*L/B   King, S.D. & Arntzen, C.J. (1989). **Plant Biotechnology**. Butterworths, 423 pp.

L/B   Mantell, S.H., Matthews, J.A. & McKee, R.A. (1987). **Principles of Plant Biotechnology: an introduction to Genetic Engineering in Plants**. Oxford, Blackwell Scientific Publication.

L/B   Mantell, S.H. & Smith, H. (Eds.) (1983). **Plant Biotechnology** SEB Seminar Series Volume 18. Cambridge, Cambridge University Press.

L/B   **Nestles Research News 1986/87**. Special Plant Biotechnology Issue.

H   Nijkamp, H.J.J., Van der Plas, L.H.W. & Van Aartrijk, J. (eds) (1990) **Progress in Plant Cellular and Molecular Biology.** Kluwer Academic Publishers. 810 pp.

L/B   OECD (1989). (Organisation for economic co-operation and development). **Biotechnology and the changing role of government**. Paris, OECD.

H   Pais, M.S.S., Mavituna, F. & Novais, J.M.(eds)(1988) **Plant Cell Biotechnology.** Proceedings of NATO Advanced Study Institute on Plant Cell Biotechnology, March 29 - April 10, 1987. Springer-Verlag. 500pp.

H   Sasson, A. (1988). **Biotechnologies and Development.** CTA Publication. 361pp.

L/B   Sasson, A. & Costarini, V. (eds) (1990) **Plant Biotechnology for Developing Countries**. CTA/FAO Publication. 368pp.

H   Sasson, A. & Costarini, V. (eds) (1991) **Biotechnologies in perspective: socioeconomic implications for developing countries.** UNESCO. 166pp.

L/B   Smith, J.E. (1988). **Biotechnology** (2nd Edn.). London, Edward Arnold.

H   Thottappilly,G., Monti,L.M., Mohan Raj, D.R. & Moore, A.W. (eds) (1992) **Biotechnology: enhancing research on tropical crops in Africa**. IITA/CTA Publication. 364pp

L/B   USDA (1986). **Yearbook of Agriculture. 'Research for Tomorrow'**. USDA Publication, 336 pp. (in periodicals section).

L/B   Zaitlin, M. (Ed.) (1985). **Biotechnology in plant science: relevance to agriculture in the eighties**. London Academic Press. (581.O).

Key journals and current awareness on plant tissue and cell culture

Bio/technology

Journal of Plant Physiology

Cell Tissue and Organ Culture

Plant Science

Plant Journal

Journal of Horticultural Science

Plant Cell Reports

Plant Physiology

Plant Cell

Theoretical and Applied Genetics

Section 3: PREPARATION

Tissue culture is a highly empirical subject with trial and error methods generally being required to arrive at technical solutions. For success, it may be necessary to vary manipulative procedures used, the methods of plant sterilisation employed, the compositions of culture media used, the conditions of incubation applied and to evaluate thoroughly a range of alternative environmental conditions under which cultures can be maintained.

Specific problems still need to be overcome, e.g. poor or inconsistent germination of somatic embryos for many plant species, internal contamination of plant tissues that becomes problematic over long periods of culture, genetic instability of some tissue culture systems and the low regeneration capacities of many monocots and woody plants that hinder attempts to develop genetic transformation systems and rapid clonal multiplication. These aspects are important plant tissue culture research topics in the 1990's.

Terminology

The term "plant tissue culture" is generally used to cover all types of aseptic plant culture. **Tissue culture** is the growth or maintenance of tissues *in vitro* in a way that allows differentiation and preservation of their architecture and/or function. **Differentiation,** i.e. the process through which cells can become specialised in structure and function, can take place in tissue cultures. In tissue explants a portion of cells retain their abilities to regenerate the whole plant. In this respect they are termed **totipotent. Aspetic techniques** are used for tissue culture work. These involve procedures which prevent the introduction of microbes into the cultures.

The following types of plant tissue cultures are generally used:

1) Culture of seeds to produce seedlings.

2) Isolated mature or immature zygotic embryos (embryo cultures).

3) Isolated plant organs (organ cultures i.e. root tips, stem sections, leaf primordia pieces, primordia or immature parts of flowers, immature fruits).

4) Tissues arising by proliferation from segments of plant organs (tissue or callus cultures). Callus is an unorganised proliferative mass of dedifferentiated plant cells and is characteristically found as a wound response at the base of stem cuttings *in vivo*. **Morphogenesis** is the process of growth and development of dedefferentiated calluses into organised structures, ie tissues and organs. Morphogenesis can take place through two main routes: **organogenesis** involves development of shoots, flowers or roots from tissue cultures, usually *de novo* from totipotent cells and **embryogenesis** which involves the initiation and development of somatic embryos from totipotent cells.

5) Isolated cells or very small cell aggregates remaining dispersed as they grow in liquid media (cell suspension cultures).

6) Callus suspension cultures are liquid culture in which cells aggregate together and do not form fine suspensions.

Main applications of plant tissue culture

***Micropropagation**: *in vitro* multiplication of plants from various tissues under aseptic conditions.

***Pathogen/Pest elimination**: carried out by culturing shoot tips or shoot apical meristems and regenerating plants from these.

***Plant breeding**: embryo rescue, somatic hybridisation (protoplast fusion), mutagenesis, dihaploid homozygous plant production (microspore and anther culture) and specific gene transfer (genetic engineering / genetic transformation)

***Germplasm conservation and transfer**: meristem culture and cryopreservation, microtubers and gel encapsulation of axillary buds and somatic embryos.

3.1. Care of mother stock plants and preferable sources of explants

* Maintain mother plants or cuttings of mother plants into a glasshouse at least two weeks before proposed culture initiation. This procedure reduces the number of microorganisms in the phylloplanes (aerial surfaces) of the plant. Spraying with appropriate insecticides and fungicides when necessary may also help to reduce microbial contaminants in the explants used for tissue culture work but such measures should be avoided in the week prior to explant removal.

* Water plants at root level only. Watering onto leaves encourages the build-up of contaminants and can lead to accumulations of micro-organisms in the leaf axils and this makes effective surface sterilisation difficult.

* Material for *in vitro* culture should not be collected from insect- or disease-damaged plants since high levels of either surface and/or endophytic microbial contamination are likely to be present in these types of tissues.

* Harvested material should be temporarily stored in a plastic bag to prevent desiccation. If the plants have wilted they should first be soaked in water to allow them to recover turgidity before being surface sterilised. Excess leaf growth should be removed from the plant material ensuring that a part of the petiole is retained to prevent hypochlorite damage of the axillary buds through cut surfaces. Do not store plant materials in plastic bags for too long in a fridge; this causes putrefaction.

* When using buds taken from underground storage organs, such tubers, rhizomes or corms, it is extremely difficult to remove all contaminating microorganisms. This problem can be reduced by growing mother plants in pots containing a clean substrate such as vermiculite or perlite under glasshouse conditions. Emerging shoot growth is then more likely to be free of excessive surface contamination than it would be if plants were growing outside under non-protected conditions. Standard surface sterilisation treatments are therefore more likely to yield explants free of surface contamination.

* Most suitable plant material for *in vitro* culture is young and actively growing, preferably produced under clean glasshouse conditions. The optimal times to take material from mother plants growing outside are in the spring/early summer (temperate regions when the new shoot growth appears) or following new flush growth at various times of the year tropical regions).

* If the plant material is in a dormant state, it is often possible to induce new growth by culturing the plant in the glasshouse. Alternatively, some plants require chilling to break bud dormancy, eg apple.

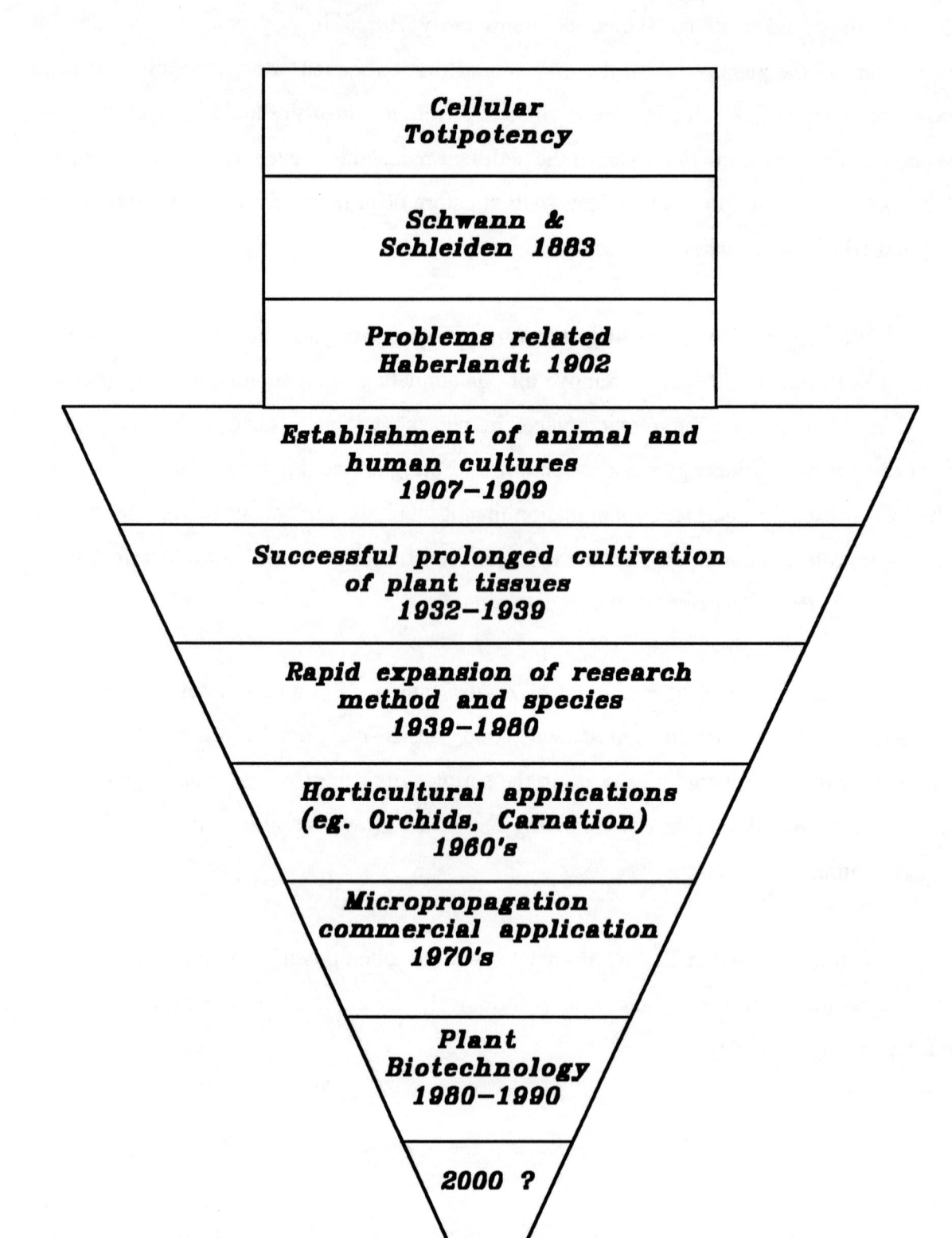

```
            ┌─────────────────┐
            │    Cellular     │
            │   Totipotency   │
            ├─────────────────┤
            │   Schwann &     │
            │ Schleiden 1883  │
            ├─────────────────┤
            │ Problems related│
            │ Haberlandt 1902 │
            └─────────────────┘
```

Establishment of animal and human cultures
1907–1909

Successful prolonged cultivation of plant tissues
1932–1939

Rapid expansion of research method and species
1939–1980

Horticultural applications (eg. Orchids, Carnation)
1960's

Micropropagation commercial application
1970's

Plant Biotechnology
1980–1990

2000 ?

Figure 1. Development of micropropagation (1838-2000)

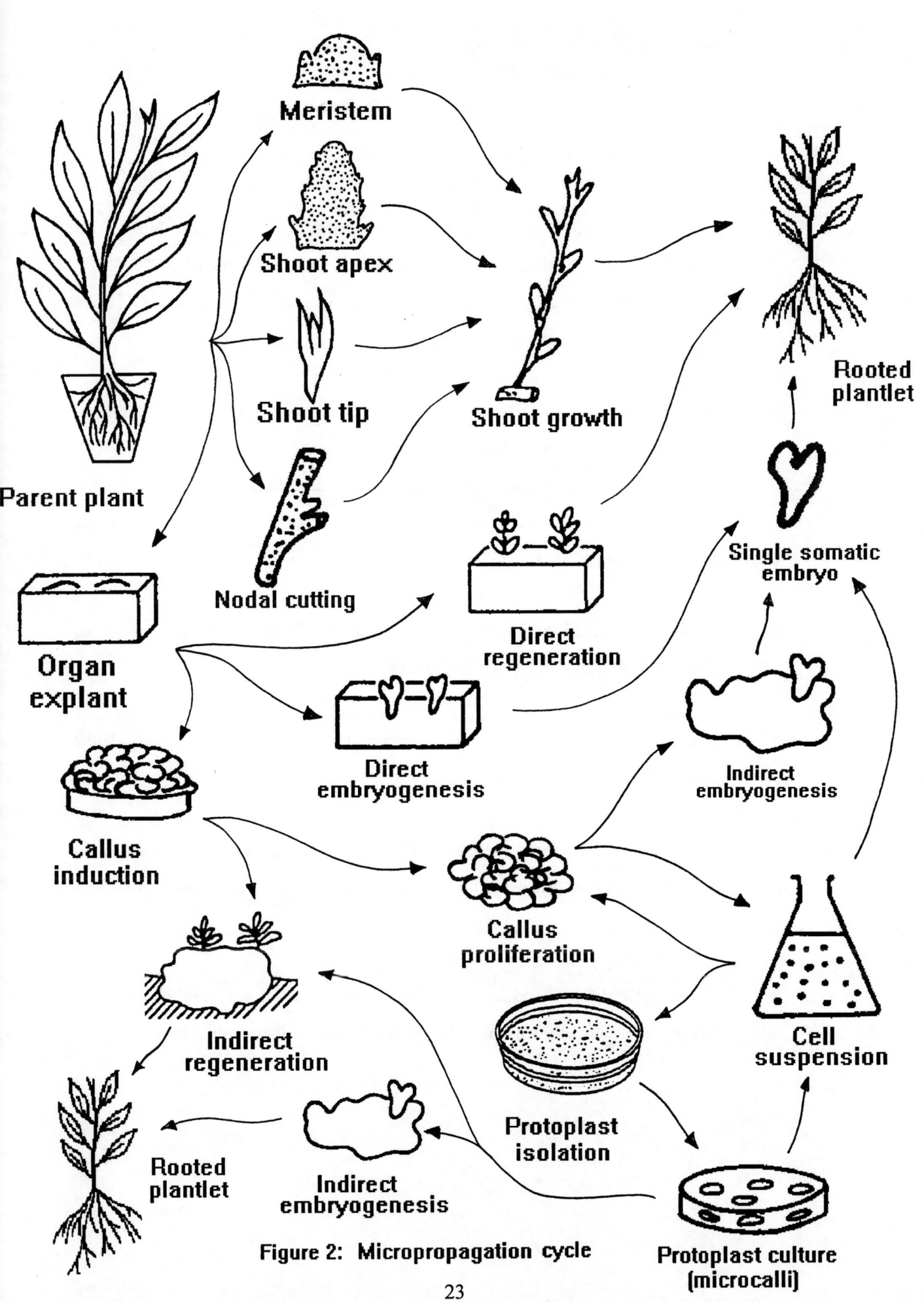

Figure 2: Micropropagation cycle

3.2. Setting up the flow cabinet

1. Wash your hands with soap and water.

2. Swab working surfaces and sides of flow cabinet with 70% (v/v) ethanol (or a more suitable domestic surface disinfectant such as "Pursue" or "Dettox") starting from the back of the flow cabinet using a sterile paper towel. Leave the paper towel on one side of the cabinet for further use.

3. Turn on Steribead glass bead steriliser or clay heater (operation light shows) and wait for temperature to rise to 200-250°C (*ca.* 15 min).

4. Turn on cabinet fan and lights and leave to run for 5-10 min before making any further preparations.

5. Prepare and position two suitable jars for 10-20 ml 70% EtOH used for instrument dipping and swabbing operations.

6. Lay out covers, instruments, tube holders and other equipment required for the culture session.

7. Sterilise instruments in a steribead / clay heater for at least 20-30 s and allow instruments to cool in sterile air flow while resting on the serrated tool-holder provided.

8. Place flasks of SDW and an empty non-sterile beaker (reservoir for receiving discarded bleach solution and rinsing water) in the flow cabinet.

3.3 Surface sterilisation of explants

1. Wash untrimmed explants under running tap water (to remove soil and dust). Following washing, these are trimmed into sections so as to fit conveniently into sterilisation flasks. At least 2cm internode tissue must be retained at each end of a stem nodal segment to prevent hypochlorite damage to buds.

* **Note**: Prolonged soaking of explants in tap water may also be advantageous in some cases to help leach away excessive phenolic and mucilaginous substances which might be inhibitory or toxic to explant tissues *in vitro*.

2. Place trimmed material into either pre-sterilised conical flasks fitted with water-tight screw-on caps or in Twyford polycarbonate sandwich boxes.

3. Make up a fresh 10% (v/v) solution of commercial bleach (10ml bleach + 90ml SDW) and add one drop (0.01%) of Tween-20 wetting agent.

4. In most cases, a short wash (10 - 60 s) in 70% ethanol is recommended to remove air bubbles from beneath leaf primordia, glandular hairs and other external structures on explants before explants are soaked in hypochlorite. Add the ethanol and decant off after 10 - 60s.

5. Add the bleach solution to the flask containing explants, seal the flask and agitate vigorously (this can be done using a mechanical shaker device or by hand) for at least 15 min. The length of time required to accomplish effective surface sterilisation will depend on the relative cleanliness of the material and upon its relative "softness". One approximate guide is to continue sterilisation treatment until *ca.* 1 mm of explant tissue has been bleached (turned white) behind each cut surface. Delicate tissues (e.g. seedlings, leaves) may only be able to withstand 5 to 10 min sterilisation.

6. After a suitable period of sterilisation the flasks containing the explants are transferred to the flow cabinet and all further procedures are carried out under strictly aseptic conditions.

7. Keep hands off all sterile parts of sterilised instruments, plates and culture vessels. Work inside the 25 cm "guideline" (ie. in the totally sterile zone) of the flow cabinet. Do not pass hands, head or any instruments over sterile surfaces. Keep all sterile surfaces clear of non-sterile objects.

8. Remove lid(s) of sterilising container(s). Keeping hands inside the "workline" decant off the bleaching solution into the reservoir beaker. Use the lids of the sterilising containers to prevent the explants from dropping into the beaker.

9. Open a bottle of sterile ROW inside the cabinet and pour sufficient (*ca.* 100ml) to cover the explants remaining in the flask. Holding the base of the flask, gently agitate the water so as to wash any hypochlorite off the explants. Discard the first wash into the beaker as previously.

10. Repeat the washing procedure until no traces of foam are left in flasks. Usually three washes suffice. Place flasks containing washed sterile explants to one side until required. In most cases, sterilised explants can be left in flasks for several hours and overnight in the fridge. Storage of these types of materials for longer than 24h is not recommended.

Additional hints

* Success rates for sterilisation tend to decrease as the growing season extends because of the increasing amounts of dirt and dust being blown onto the developing plant surfaces. As aging proceeds more microbes are likely to become embedded in the waxy layers of cuticle covering aerial parts of plants. A brief pre-treatment of explants with absolute alcohol or chloroform may help to dissolve the surface waxes and so expose any embedded microbes to surface sterilisation treatments.

* If the standard sterilisation procedure is proving unsuccessful, then either the standard sterilisation process can be repeated several times (increasing the levels of bleach or the length of exposure time as the process is repeated) or stronger sterilants can be used. For instance, a solution of 0.1% mercuric chloride can be used and may be used to sterilise explant tissues obtained from field-grown mother plants or from material which are in contact with the soil eg. tubers, bulbs, corms or stump shoots.

> **WARNING - MERCURIC CHLORIDE IS HIGHLY TOXIC TO THE ENVIRONMENT AND THAT INCLUDES YOU AND OTHERS IN THE LAB!**

Other types of strong sterilants used for plant tissue culture work are hydrogen peroxide, potassium permanganate and a range of ammonium-containing medical sterilants (for further details see George and Sherington, 1984)

* Pre-washing of explants with sterile distilled water or under running tap water can sometimes be of benefit because this treatment dilutes down the microbial populations present.

* **Antibiotic** supplements in the first culture medium can facilitate the reduction in populations of both any residual bacteria remaining after sterilisation treatments and endophytic microbes. The sterile tissue should then be transferred to antibiotic-free medium at the first subculture.

Section 4: PRACTICAL EXERCISES

Exercise 1: Introduction to various procedures

4.1.1. Medium preparation

1. See UAPS Manual Volume 2 (Section III Part A5 to A6) for all procedures relating to the preparation of a plant tissue culture medium.

2. Make up stock solutions for Murashige and Skoog (1962) (MS) medium as given in Table 1. Also prepare a series of stock solutions for relevant growth regulators.

3. Stock solutions are generally used in those cases where only small quantities of a substance are required in a medium recipe. For instance, in MS medium, $MnSO_4.4H_2O$ is required at a level of 22.3 mg/l (ie 0.0223 g/l). This quantity is far too small to measure out accurately each time MS medium is prepared. It is more convenient (and more accurate!) to weigh a quantity which is 100 x of that required, dissolving the compound in water and then dispensing an appropriate volume of stock solution containing the desired quantity of compound into the medium. Hence for a 100 x stock solution of $MnSO_4.4H_2O$, 2.23g $MnSO_4.4H_2O$ is dissolved in 1000ml of ROW. However, only 0.0223 g/l is required in the medium. Therefore, the amount (volume) of stock solution required = 0.0223/2.23x1000; 10ml stock solution is required for each litre of medium prepared.

*This calculation can also be expressed in the following form:

$$V_1 C_1 = V_2 C_2$$

Where: V_2 = Volume of the stock solution (normally 1l)
V_1 = Volume of stock solution required for 1l of medium
C_1 = Concentration of stock solution (mg/l or Molar)
C_2 = Concentration of compound required in 1l medium (mg/l or Molar)

*Dissolving media constituents

i) Macronutrients and micronutrients dissolve in water but $Mg.SO_4.6H_2O$, and KH_2PO_4 must be added and dissolved slowly and in the given sequence to prevent precipitation. $CaCl_2.2H_2O$ is dissolved separately and an individual stock solution prepared. Each salt must be completely dissolved before the next is added.

ii) Vitamins are all soluble in water.

iii) Growth regulators: cytokinins are readily soluble in HCl. Indole auxins and NAA can be dissolved in NaOH. 2,4-D can be dissolved in ethanol but the use of dimethyl sulphoxide (DMSO) is sometimes used. **Warning**: DMSO has toxic effects on plants.

Important points to remember

If the stock solution is too concentrated, recrystallisation of the compound may occur at low temperatures, eg when stored in a fridge. Most stock solutions are kept in the cold room (4°C) for a longer storage life. If the stock solution is too dilute, then large volumes may be needed to be added to the medium. This can lead to problems of over addition of liquids, thus final volumes can be easily exceeded.

4. To the medium, add growth regulators (as required) and sucrose (2 - 4%), stir with a teflon-coated magnetic bar until all the constituents are dissolved. Adjust medium pH to 5.7 - 5.8 (using either 0.1N HCl or 0.1M NaOH) and then add the required amount of Technical Grade 3 agar (8 g/l).

5. Heat the stirred medium to dissolve the agar and dispense appropriate volume into culture vessels (Magenta top tubes or jars). Usually 5-7 ml medium are dispensed into each soda glass tube and 15-30 ml are dispensed into each glass jars (depending on size).

6. Place lids on the culture vessels and wrap the racks of tubes or jars in greaseproof paper before autoclaving. Label each pack of tubes with your name, the current date and the type of medium used. Stick a small piece of autoclave tape onto the parcel and place on the appropriate shelves in the autoclave room.

Table 1. Preparation of MS medium using stock solutions

Component	Stock solution concentration	Stock solution volume per litre medium	Final medium concentration (mg/l)
A) Macronutrients g/1000ml (20x) NH_4NO_3 KNO_3 $MgSO_4.7H_2O$ KH_2PO_4	33 38 7.4 3.4	50ml	1650 1900 370 170
B) Macronutrients g/1000ml (20x) B) $CaCl_2.2H_2O$	8.8		440
C) Micronutrients mg/1000ml (200x) **$MnSO_4.4H_2O$** $ZnSO_4.7H_2O$ H_3BO_3 KI $Na_2MoO_4.2H_2O$ $CuSO_4.5H_2O$ $CoCl_2.6H_2O$	**4460** 1720 1240 166 50 5 5	5ml	**22.3** 8.6 6.2 0.83 0.25 0.025 0.025
D) Iron compound mg/100ml (20x) $Na_2EDTA.2H_2O$ $FeSO_4.7H_2O$	672 556	5ml	33.6 27.8
E) Vitamins mg/100ml (100x) myo-inositol	1000	10ml	100
mg/100ml 100x **F) Organic** Glycine Nicotinic acid Pyridoxine HCl Thiamine HCl	20 5 5 1	10ml	2.0 0.50 0.50 0.01

4.1.2. Calculation exercise

*1 Molar = Molecular Weight (MW) of a compound in 1 litre
(Moles = Molecular weight expressed in g)

eg MW of 2,4-D = 221

1M 2,4-D = 221g 2,4-D in 1 litre

1×10^{-3} M, ie 1mM = 221mg in 1 litre (see Table 2).

A 2,4-D stock solution of 1×10^{-3}M therefore contains 221 mg 2,4-D in 100mls. If the final concentration required in one litre of medium is 1×10^{-5}M, then using the equation:

$$V_1 M_1 = V_2 M_2$$

Where:

V_1 = Volume of stock solution

M_1 = Molarity of the stock solution

V_2 = Volume of stock solution containing the required amount of dissolved compound

M_2 = Molarity of required compound in 1l of medium.

Therefore, to give a final concentration of 2,4-D in 1 litre medium as 1×10^{-5}M, 10ml stock solution must be added and the final volume then made up.

Table 2: Interconversion between mass/volume (g/l; mg/l; µg/l and molarity (M) measurements

g/l	mg/l	µg/l	M	M	mM	µM
221.0	221000	-	$= 10^0$	1	1000	1000000
22.1	22100	-	$= 10^{-1}$	0.1	100	100000
2.21	2210	-	$= 10^{-2}$	0.01	10	10000
0.221	221	221000	$= 10^{-3}$	0.001	1	1000
0.0221	22.1	22100	$= 10^{-4}$	0.0001	0.1	100
0.0022	2.21	2210	$= 10^{-5}$	0.00001	0.01	10
-	0.221	221	$= 10^{-6}$	-	0.001	1
-	0.0221	22.1	$= 10^{-7}$	-	0.0001	0.1

4.1.3. Surface sterilisation exercise

At least two plant species will be used to test the effectiveness of different surface sterilisation procedures. Both plants may be difficult to sterilise and a variety of sterilisation methods will be tried to test which is the most effective.

Treatments

A   Control (15 min agitation in a 10% (v/v) bleach solution)

Explants will be passed through normal procedures except that:

B   2 h in running water prior to sterilisation

C   1 min in EtOH (70%) prior to sterilisation

D   Four 10 min washes (on the shaker) in sterilised distilled water prior to sterilisation.

E   1 min in EtOH (70%) followed by four 10 min washes in SDW prior to sterilisation in 10% bleach for 15 min.

F   As for "E" except 100% bleach used.

G   15 min vacuum sterilisation in bleach followed by 15 min normal bleach sterilisation.

H   5 min vacuum sterilisation in bleach followed by 5 min normal bleach sterilisation.

Materials

* Plant material e.g. carnation, anthurium, *Solanum*, carrot seeds. Washed and cleaned in running water for 5 min (control).
* One vacuum flask with air tight stopper.
* One sterile 500ml screw-top flask or bottle with water-tight lid.
* 2 l of sterile distilled water (autoclaved for 40 min)
* 200ml 10% (v/v) bleach ("Brobat") solution containing one drop (0.01%) of wetting agent (Tween 20).
* One pair of dissection scissors or razor blade for cutting up material.
* Two scalpel handles with new blades.
* Two pairs of forceps (one long and one short).

* One non-sterile reservoir beaker for discarded bleach and water.
* Two boxes of polypropylene discs (a large and a small size)

Procedures

(A) *In the Preparation Room*

NB. Do not forget to vary Treatment B as above.

1. Prepare explants (segments) from the collected plant material.
2. Wash under running tap water (standard 5 min, Treatment B for 2h) to remove any superficial dirt from the explants.
3. Cut tissues up into sizes which will fit into the sterilisation bottle and if necessary leave to soak in running water for an hour or two to leach out exudate and to eliminate contaminants.
4. Remove excess leaf laminae but do not forget to retain the petioles in order to prevent infiltration of the bleach solution into the axillary buds.
5. Now the plant material is ready to be taken into the lamina air flow cabinet for sterilisation (in a clean container).

(B) *In the Transfer Room*

1. All operations from here on are conducted under aseptic conditions and as indicated previously with modifications according to the various treatments (C-H). Plan to have at least 10 explants per treatment and do not forget the Controls.
2. Draw a flow chart indicating the layout of the various treatments and controls. This approach will be discussed during the practical sessions.
3. Proceed with the preparation of explants using the forceps, scalpels and polypropylene disc (see Section 3.4).
4. Transfer the explants to fresh medium and incubate for one week and determine the proportions of explants free of contamination. Also make notes on the appearances and development of the explants (see Section 4.3 for layout of data sheet).

4.1.4. Dissection and culturing of explants

1. Prepare sterile polypropylene circle sheets on the cutting surface (tile or glass plate).

2. Using sterile long tweezers, transfer one of the sterilised pieces of explant material onto the polypropylene sheet. Resterilise tweezers and allow to cool. This will be demonstrated to you.

3. Using cool short tweezers and scalpel, remove all bleach-damaged tissues. To cut material hold the tissue with the tweezers very close to the point of cutting to prevent tearing. Cut with a sawing motion especially on soft tissues. Pressing down with the scalpel to cut causes extensive tissue bruising damage which stimulates excessive ethylene production thus retarding growth *in vitro*. **Work carefully with delicate plant tissues!**

4. Prepare no more than six large leafy explants at each preparation before transferring trimmed explants into culture vessels so as to avoid excessive desiccation of tissues. Explants without leaves do not dry out as quickly as those with leaves. Small or leafy explants can be stored longer if kept in sterilised ROW or in liquid medium before dissecting and transfering into culture vessels.

5. Hold the pair of long-nosed tweezers with one hand and a culture tube with the other. Remove the lid of the tube, using the tweezer hand and hold it. Using the tweezers, pick up an explant and transfer it into the culture tube. Seal the tube with a polypropylene square and an elastic band. Resterilise the tweezers. The inoculation process will be demonstrated to you.

6. Repeat the sequence until all explants have been placed into a culture vessel.

Exercise 2: Establishing callus and cell suspension cultures

4.2.1. Callus initiation

Callus (plural: calluses or calli) is formed naturally from rapid cell divisions which occur at or near the site of wounded cells at the cut surfaces of plant tissue. This wounding response assists it seal the cut surfaces and protects the underlying cells from dessication and infection by external microbes. Larger amounts of this type of tissue can be induced when wounded explants are placed on a medium containing growth regulators, such as 2,4-D, NAA and cytokinins. In the absence of physical constriction imposed by epidermal tissues, plant cells grow in an disorganised manner. By contrast, cell divisions which occur within an organised organ such as a bud are more precisely controlled by the surrounding tissue environments. Consequently, new divisions which occur within a tissue usually proceed in an orderly fashion. Callus cells are large with thin cell walls and are not highly cytoplasmic. They do not have strong cell-to-cell connections and can be very friable. However, they all contain a nucleus which contains the genetic material required to form a whole new plant, ie they are potentially **"totipotent"**. Given the right chemical and physiological signals, certain of the callus cells can switch to morphogenic pathways of development and new meristems can be organised to form primary nodules (in monocots particularly), roots, shoots, flowers or embryos. In practice, calluses contain a range of different cell types some of which are totipotent and the degrees of dedifferentiation / organisation which they exhibit may vary depending on the origins of explants and culture conditions employed by the experimenter. Unlike zygotic embryos, the embryos formed from cell suspensions and calluses are termed "somatic embryos" and should generate clones of the mother plant. However, in practice the clonal status of somatic embryos should always be tested thoroughly because of the risks of somaclonal variation occuring during *in vitro* culture when callus growth phases can be prolonged and genetic deviations tend to increase with increasing periods of culture.

The growth of callus and cell suspension cultures follows an exponential growth curve. A lag phase (where cells become meristematic and prepare for division) is followed by an exponential phase in which cell division is prolific. A plateau in growth of cultures is reached when nutrients become limiting and cell divisions cease. To obtain embryogenic callus, it is essential to continue subculturing the calluses in order to maintain active cell divisions in the cultures. This is because actively growing cells respond effectively to external hormonal/nutrient signals. Not all callus cells will be embryogenic, although regular subculture should increase the percentage totipotent cells present. Experience is usually required to identify the visual appearances, ie colour and texture, of calluses which are embryogenic. Embryogenic callus usually develops on an explant when explants are placed on a medium containing either low levels of 2,4-D or no hormones at all (in the case of nucellar tissues of polyembryogenic species, eg citrus). Embryo induction on 2,4-D medium is stimulated by a series of subcultures on media with either reduced auxin level or with no auxin so that later stages of embryo development and plantlet production are stimulated.

Calluses can be subcultured in a liquid medium in conical flasks which are placed on a rotating platform. As the calluses proliferate, cells are sloughed off and suspensions of cells and small cell aggregates are formed. To prevent the cells from clumping together, it is necessary to filter / sieve the cell suspensions regularly through sterile gauze into fresh culture medium. This is usually most conveniently done on the occasion of each subculture.

Callus and cell suspensions are often used for mutation work since individual cells can be mutated and selected through use of plating techniques. Plant breeders can also make use of the **somaclonal variation** that may occur before culture in populations of cells within tissues *in situ* in the mother plant or in individual cells cultured separately *in vitro*. The levels of variability can be increased substantially (by several orders of magnitude) through the use of physical or chemical mutagens.

Aim of experiment
* To induce calluses from different explants of carrot (*Daucus carota*) -cambial tissues in the storage root, potato (*Solanum tuberosum*) - tuber discs, and tobacco (*Nicotiana tabacum*) - leaf, stem segments.
* To generate cell suspensions from calluses of these species.

Materials

* Large, healthy tap root or sterile seedlings (previous experiment) of carrot (*Daucus carota*), a large healthy tuber of the potato (*Solanum tuberosum*) or healthy plants of tobacco (*Nicotiana tabacum*).
* Vegetable peeler
* 2 x 500ml watertight sealed flask or bottles containing SDW.
* 40ml bleach concentrate + 2 drops Tween 20 wetting agent.
* 1 x 600ml beaker (not sterile).
* 2 x 500ml bottles of SDW.
* Stainless steel cork borer (No. 2) containing metal or glass rod. Enclose borer in a test tube or bottle and enclose in a layer of foil. Not required if working with tobacco.
* Two pairs of forceps
* Two scalpels with blades (No. 4 handle + No. 22 blade).
* 30 x 2.5cm diam. culture tubes containing 10ml MS medium. Make up 300ml MS medium supplemented with sucrose (3% w/v) + 20µM 2,4-D or 20µM NAA and 4 µM BAP + 8 g/l agar. pH 5.7 - 5.8.
* One box of sterile, circular polypropylene sheets for dissections.
* One box of sterile, square polypropylene covers for sealing tubes.
* One tile or glass plate
* Two sterile 9cm petri dishes to hold sterile water and prepared explants.
* One beaker of tap water

Procedures

1. Wash the tap root/tuber/stem material under running tap water in the dirty laboratory.

2. Remove the external 1-2mm of tissue with a vegetable peeler. The tobacco stems must be cut into 2 node sections and placed into a beaker of tap water (use *in vitro* tobacco plants if available). All injured tissues must be removed.

3. Cut the carrot and potato into 10mm thick slices and place them into a beaker of tap water.

4. Wash your hands

5. Prepare the lamina air flow cabinet first by swabbing with 70% ethanol with a sterile tissue. Turn the fan and the steribead steriliser on.

6. Allow the steribead steriliser to attain 250°C and then place all scalpels and tweezers, after dipping in 70% EtOH, into it for at least 40s. Remove tools from steriliser and allow to cool down on the serrated tool holder.

7. Place the plant material into the bottle containing 300ml of the bleach prepared. Seal the container and agitate for ten minutes on a vibrator device or by hand.

8. Following the 10 min sterilisation period, empty the bleach solution from the flask into a non-sterile beaker and pour c. 200ml SDW into the flask. Agitate for c. 20s and discard the wash water into the same beaker. Repeat this sequence two more times.

9. When all traces of bleach have been removed store the carrot sections in c.100ml of SDW to prevent desiccation.

10. Place a polypropylene disk onto the tile and using sterile tweezers transfer a carrot/potato disk or tobacco stem section onto it.

11. Holding the carrot/potato disk firmly with the tweezers, use the borer to cut out a section of tissue which includes the vascular ring (containing the main meristem tissues).

12. Lift the slice, with the borer still inserted in it, and hold directly over the SDW in the partially opened Petri dish. Gently exert pressure on the metal rod to eject the cylinder. Prepare a maximum number of cylinders out of each disk.

13. Tobacco stems, after cutting away bleach damaged sections, are cut up into internodal sections *c.* 0.5cm long stored in sterile water until required for culture.

14. Place a fresh sterile polypropylene disk onto the tile, transfer some tissue cylinders onto the dish, and then using the tweezers and scalpel remove the bleach damaged parts of the carrot and potato cylinders.

15. Slice the remaining cylinder into three pieces approximately 2mm in length.

16. If using sterile carrot seedlings simply cut into 3mm sections and place onto solid or into liquid medium. Place at least 1g tissue into 5-10ml liquid medium.

17. Transfer the potato, carrot or tobacco explants to the surface of the culture medium and seal the tube with a polypropylene square.

18. Incubate the cultures in the dark at 25°C for about 1 month to allow the growth of callus.

19. Count the number of contaminated tubes, discard them in the prescribed way and express the number of contaminated cultures as a percentage of the total number of explants.

4.2.2. Cell Suspension initiation

The initiation of a cell suspension culture requires a relatively large amount of callus to serve as the inoculum (approximately 2-3g / 100ml medium). When the plant material is first placed in the medium, there is an initial lag period prior to any sign of cell division. This is followed by an exponential rise in cell number, and a linear increase in the cell population before there is a gradual deceleration in the division rate. Finally, the cells enter a stationary or non-dividing stage. Sub-culture should be carried out during the stationary phase to maintain culture viability.

Materials

* Actively growing callus cultures of *Daucus carota*, *Nicotiana tabacum* or *Solanum tuberosum* along with other species when ever available.
* Two pairs of forceps
* Two scalpels
* Container of polypropylene disks (larger sheets)
* Container of polypropylene squares
* 5 x flasks each containing 25ml liquid MS medium + 4µM 2,4-D + 30g/l sucrose.
* 5 x empty flasks sealed with foil
* 0.1mm diameter stainless steel or nylon mesh filter (sterile)
* 10 x sterile disposable wide-bore pipettes
* Rubber pipette filler
* One sterile funnel
* One 50ml beaker (not sterile)

Procedure

Remove callus from explants cultured in the previous experiment. The callus may be hard to remove so use sharp scalpel blades. Remove all differentiated tissues so that only actively growing callus (undifferentiated tissues) remains. Brown callus should be discarded.

1. Place about c. 500 - 750mg of callus into each culture flask. Seal flasks with Nescofilm foil.

2. Agitate the cultures at 100 rpm on a shaker apparatus at a temperature of 25-27°C in dark.

3. Turbidity of the solution indicates that contamination has probably occurred.

4. First subculture after 7-10 days. Filter the suspension culture through the mesh filter, to remove large clumps of tissue and cell clusters, into a clean sterile flask.

5. Remove a 1ml aliquot of the cell suspension culture with a sterile pipette and discharge the contents into a small beaker. Place c. 0.1ml of the suspension onto a glass slide and examine under a microscope.

6. After a further 7 to 10 days culture, subculture by adding equal volume of fresh medium of the same composition and dividing the contents into 2 equal volumes in separate sterile flasks. Thereafter, subculture every 14 to 21 days depending on the density of the suspension culture and refilter the suspension each time to separate large callus clusters from smaller cell colony or single cells.

Exercise 3: Micropropagation: Stages I, II and III

4.3.1. Meristem and shoot apex cultures

The apical meristem refers only to the region of the shoot apex above the distal to youngest leaf primordium, whereas the shoot apex refers to the apical meristem plus a few subjacent (in a proximal direction) leaf primordia. The excision of apical meristem (80μm long) is time consuming and they have a low survival rate, however, they are important in the development of pathogen-free stock such as virus particles and other systemic pathogens do not usually penetrate the young actively dividing the meristem. Inoculation of plants at 35-38°C for 2-3 weeks may enable the shoot tip to outgrow the spread of virus particles and hence be termed "virus-free". The use of this term should be discouraged because it is never possible to be absolutely certain that a plant stock is completely free of virus. It is more preferable to use the term "virus-tested" since this infers that the plant stock in question is free of known viruses detected by diagnostic tests that are specific for known viruses. In other words, the material can only be guaranteed to be free of specific viruses. Chemical anti-viral treatments can also be used separately or in combination with hot air therapy to eliminate particularly infectious viruses such as potyviruses.

Shoot apex culture is frequently used for the clonal multiplication of certain plants. The method of micropropagation with shoot tips is based on the cytokinin-induced outgrowth of bud primordia, each of which produces a miniature shoot.

Factors influencing successful shoot apex culture at various stages:

Stage I: *Establishment of aseptic cultures*
* Choice of suitable explant-shoot tips and buds excised from healthy and actively growing herbaceous plants are generally ideal material for multiple shoot production.
* The larger the size of the tip explant, the more rapid the growth and the greater the rates of survival.

* A procedure not involving a callus phase is preferable.

* Murashige and Skoog (1962) medium is satisfactory in most cases.

* The cultures can be grown on agar, on filter-paper bridges, polyurethane sponges or on Sorbarods using liquid medium.

* Light intensity, photoperiod (e.g. 16h day) and culture temperature all play a role in determining the success or failure of plant regeneration from meristem explants.

* Choice and concentration of growth regulators: cytokinins such as kinetin, benzyladenine, IPA, thidiazuron or zeatin may be used alone or in conjunction with IAA, NAA or IBA. The auxin 2,4-D is unsatisfactory, since it tends to stimulate greater amounts of callus formation.

Stage II: Multiplication of propagules by repeated subcultures on cytokinin containing medium

* Typically, although not always, the same medium and environmental conditions are used for both stages I & II.

Stage III: Preparation of microplants for establishment ex vitro

* Involves the development of a root system, hardening the young plants to avoid excessive moisture stress, increasing resistance to certain pathogens, and conversion of the plants to a photoautotrophic state.

* Rooting may be facilitated by treating explants with low levels (1-5μM) of NAA or IBA, but treatment must be limited to a brief period of time because prolonged auxin treatment can inhibit root elongation and growth. Auxin is required for the root induction stages only.

Materials

* Shoot-apical material from *Solanum tuberosum*, *Pelargonium* sp. (geranium), *Dianthus* sp. (carnation), *Vitis* sp. (grapevine) or other plant species.
* Two scalpels
* Two pairs of forceps
* Two new razor blades (sterile)
* 12 filter-paper bridges prepared from Whatman No. 1 filter paper (9 x 90mm strips).
* 12 Sorbarods
* 12 x 3-inch "Poly-Top" tubes, each containing a filter-paper bridge and 3ml MS medium supplemented with 2 mg/l cytomix and 2% sucrose.
* 12 x 3 inch "Poly-Top" tubes, each containing a Sorbarod and 8ml MS medium supplemented with 2 mg/l cytomix as above and 2% sucrose.
* 12 x 3 inch "Poly-Top" tubes, each containing 10ml MS, supplemented with 2 mg/l cytomix, 2% sucrose and agar (8% w/v).
* All chemicals, water and containers required for sterilisation - see Section 4.1.3.
* 9cm diameter petri-dish containing 5ml sterile distilled water.
* Binocular dissection microscope (50 x magnification) and a piece of graph paper ($4cm^2$).
* Sterile 5ml pipettes
* Pack of sterile petri dishes (9cm)
* 30 sheets of sterile filter paper

Procedure

1. Insert a filter paper bridge folded in the shape of the letter "M", or a Sorbarod into each culture tube. Add the correct volume of medium and seal the tube. Wrap up in greaseproof paper and autoclave.

2. Apical shoots 1cm long are removed from the plant and placed in the bleach solution for c. 15 min. All subsequent procedures are conducted aseptically.

3. Wash off bleach by rinsing shoots several times in SDW.

4. Transfer shoots to a petri dish containing a sheet of filter paper for surgical removal of the shoot apex. Moisten the filter paper with a few drops of sterile water to prevent shoot tip desiccation.

5. The terminal growing tip, 0.5-1mm in length, is carefully excised with the aid of a 10x to 20x binocular dissection microscope. Observations will be made on the optimal size of the explant to use for micropropagation (this will be assessed using graph paper placed below the glass cutting surface).

6. Each shoot tip is transferred to the filter-paper bridge, sorbarod or agar, sealed and incubated in a plant growth chamber.

7. After a few weeks the first signs of shoot proliferation should be apparent. When the shoots are well developed they are excised and transferred to root-induction medium. *See "Rooting section" (3.6.1.). Rooting medium contains only an auxin and root initiation should only take 4 - 6 weeks* (can vary among different plant species).

8. Once an adequate root system has developed the plantlets can be transferred to glasshouse conditions. See "Weaning" section below.

4.3.2. Rooting exercise

Some easily rooted plants species produce new roots during stages I and II. A separate rooting stage is therefore unnecessary, and rooted plantlets can be moved directly into the external environment for hardening off. If shoots can be given a highly concentrated auxin dip and placed directly into compost for rooting, then the extension of expensive *in vitro* methods need not be utilised. However, some plants do not root easily, *in vivo* and *in vitro* methods are necessary. Frequently, root formation is inhibited by the cytokinins used to induce shoot multiplication (Stage II), so that shoots do not form roots *in vitro* until cultured on a medium containing auxin alone or none at all.

Auxins used for rooting include IAA (1 - 100µM), NAA (0.25 - 10µM) and IBA (2.5 - 15µM). The explants are normally cultured on media containing these auxins until rooting occurs. An alternative is to utilise a concentrated auxin dip (>250µM) into which the shoot bases are soaked for 20s to 18h depending on the auxin concentration used. Treated shoots are then transferred to growth regulator-free media. Environmental conditions such as temperature, and the presence or absence of sugar play a role in the induction of rooting.

Materials

* Healthy proliferated shoots of a suitable woody species (eg jackfruit, apple, citrus) *ca.* 2cm long consisting of 2-3 nodes
* Two scalpels with new blades
* Two pairs of forceps
* Twyford box containing plastic mesh and sufficient NAA solution (2.5mM) to cover the mesh
* 30 x 3cm tubes containing MS medium without growth regulators, sucrose (30g/l), agar (8g/l), pH 5.7 - 5.8
* 30 x tubes containing MS medium + 2mg/l IBA
* One box each of polypropylene disks and squares

* One sterile beaker
* 1litre SDW

Procedures

1. Under aseptic conditions, excise suitable elongated shoots from the proliferating cluster. Place the cuttings into the beaker filled with SDW to prevent desiccation.

2. Remove leaves from the base of each shoot.

3. Dip half the number of shoots into the 2.5mM NAA bath, leave for 10 min and place the remainder of the cuttings into the tubes containing medium supplemented with 10μM IBA. Seal the tubes with polypropylene sheets.

4. After 10 min transfer the dipped shoots onto medium without growth regulators. Seal the tubes.

5. Place 10 tubes of each treatment in a different environment (25°C/16h day; 19°C/16h day, or 29°C/12h day). Every week note the appearances of the cuttings.

Exercise 4: Micropropagation : Stage IV

4.4.1. Weaning of micropropagated plantlets/acclimatization

Plants growing *in vitro* are subject to:

* High humidity and plants growing under these conditions do not normally have a well - developed waxy epidermis or functioning stomata.

* Low light levels and being exposed to bright outside light may damage the plants.

* Use of sugars in the medium encourages growth of microbial organisms as soon as plants are exposed to the external environment.

* Sterile conditions. Microorganisms can easily colonise the surfaces of an *in vitro* plant and thus reduces its chances of survival.

In view of these factors, there is a need for a transitional period between culturing plants *in vitro* and establishing them under non-sterile, environmentally stressful conditions during which humidity is lowered and light levels are increased gradually. This is known as the acclimatization process during weaning of *in vitro* plantlets.

Notes:

In order to reduce the humidity gradually, rooted microcuttings (plantlets) can be placed in a humid chamber (e.g. fogging unit or mist bench) or propagating trays with plastic lids in which there are adjustable wents. Alternatively, sterile tissue papers can be used to cap the culture vessels a few weeks prior to weaning. These conditions encourage development of functional stomata and a more substantial cuticle layer (wax) which prevents water loss from leaves.

Procedures

1. Prepare a range of suitable weaning substrates such as 1:1 perlite/vermiculite or various combinations of heat compost and vermiculite. Remember that an ideal weaning compost is one that has a good water retention capacity, but still allows free drainage. Some fast growing herbaceous plants require higher levels of nutrients than slower growing hardwood plantlets. Slow release fertilisers can be used once plants are growing on successfully. Bottom heating of the weaning bed can be beneficial also for rooting.

2. Before planting, remove agar medium from microplants to discourage the rapid growth of fungus and bacteria on the sugar containing nutrient mixtures.

3. Immerse plantlets in a fungicidal dip, e.g. Benlate or Captan, for a few seconds prior to planting to give protection against fungi.

4. Handle the microplants delicately and insert the shoots into small depressions in the weaning substrate using a suitable stick or spatula.

5. Place planted containers in either misting, fogging or propagator trays. Shade containers with muslin to protect the delicate plantlets from direct sunlight during the first week or so following weaning.

6. Record the progress of plant growth noting the various positions from which new shoots emerge. For instance, do the existing shoots of the microplants contribute to the structure of the new plants or is it newly emerging shoots which make up the new plants?

7. Once new leaves are being formed, the microplants can be transferred to non-protected nursery bed conditions that are used in traditional seed propagation approaches.

> **Exercise 5: Morphogenesis.**

4.5.1. Organogenesis (direct/indirect).

Organogenesis is the process whereby organs such as roots, shoots and flowers develop asynchronously. Adventitious organ development refers to those cases where organs are formed in positions not previously found in the whole plant. Regeneration generally refers to production of plantlets (with roots and shoots) either from an intact tissue, eg lamina pieces, petiole, midrib, hypocotyl, cotyledon explants, or from undifferentiated tissues (eg calluses). Regeneration can be through embryogenesis as well (this is discussed under a separate section since this process is distinct from organogenesis because shoot and root organs develop synchronously). Organogenesis is usually controlled by a balance between cytokinin and auxin. A relatively high auxin:cytokinin ratio induces root formation whereas a low ratio favours shoot production.

Materials

* Leaves (both young and old) excised from mature plants, eg african violet, coleus, tobacco and *Solanum aviculare*.
* Chemicals, water and bottles required for surface sterilisation of leaves.
* Two scalpels and two pairs of forceps
* One container of polypropylene disks
* One container of polypropylene squares
* Three batches of MS medium supplemented with 3% sucrose and the following combination of growth regulators as specified below:

    Batch 1 - no growth regulators

    Batch 2 - 1µM NAA + 10µM BAP

    Batch 3 - 10µM NAA + 1µM BAP

    Media solidified with 0.8% (w/v) agar and pH adjusted to 5.7-5.8.
* 500 ml sterile distilled water.

Procedures

1. Remove the petioles from all leaves. Wash the leaves in cool soapy water and then wash them in running tap water.

2. Dip leaves in ethanol (70% v/v) for 10s and rinse in sterile ROW. Surface sterilise in the bleach solution for 10 min but avoid serious damage to the leaves. The two-step sterilisation procedure is necessary because of the many epidermal hairs on the leaves. Rinse three times in SDW.

3. Use *in vitro* plantlets as starting material by passes this sterilisation process.

4. Transfer a leaf blade to a polypropylene disk. The blade tissue most effective in organogenesis is located in the central part. Slice the blade into squares (10-12mm square), ensuring each explant contains a portion of midvein. Place explants individually in culture tubes in with upper (adaxial) surfaces down position on medium.

5. Place cultures in growth chamber maintained at 25°C with 16h photoperiod.

6. After shoots have appeared they can be aseptically subdivided and further multiplied or rooted. Rooting is promoted by transferring to a medium without growth regulators or the plantlets can simply be planted out after which rooting will occur.

4.5.2 Somatic embryogenesis

Somatic embryos are initiated in callus from superficial clumps of cells associated with highly vacuolated cells that do not take part in embryogenesis. Embryoid-forming cells are densely cytoplasmic, have large starch grains, and a relatively large nucleus with dark staining nucleolus. Each developing embryoid passes through the sequential stages of embryo formation (i.e. globular, heart-shape (not seen in monocots), and torpedo shape).

Two types of media are required for the induction of somatic embryogenesis:

1. The induction of somatic embryogenesis occurs on media supplemented with auxin (normally 2,4-D or NAA).

2. The proliferation of embryo occurs on media with a lower concentration of auxin or without auxins.

The most embryogenic callus and cell suspension are those that are frequently subcultured to ensure fresh cell production. Older callus cells lose the embryogenic ability after a period of time.

Materials

* Suspension culture of carrot, tobacco or yam growing in an MS medium supplemented with 4µM 2,4-D and 3% (w/v) sucrose.

---

* **Medium A: 200ml MS medium**
 + 1µM zeatin + 0.5µM 2,4-D (4.5µM) + 2% (w/v) sucrose
 + 1% (w/v) agar

Medium B: 200ml MS medium
+ 1μM zeatin + 2% (w/v) sucrose
+ 1% (w/v) agar

Medium C: 200ml MS medium
+ 1μM kinetin + 2% (w/v) sucrose
+ filter paper bridges

* 15 sterile disposable petri dishes (9cm diam.)
* One pack of sterile 2ml pipettes
* Two pairs forceps and two scalpels

Procedure

1. When the autoclaved MS medium (A and B) are hand-hot, five petri-dishes of each medium type must be poured under aseptic conditions. Allow the medium to cool.

2. 2ml carrot suspension culture must be added by pipette to the surface of the medium in the Petri-dishes. The dishes are sealed with Nesco-film and inoculated at 25°C in the dark for 2 - 3 weeks after which time somatic embryos should have developed.

3. After the carrot cultures have reached late torpedo stage they can be transferred onto sorbarods in tubes containing about 7ml of liquid medium or onto solid medium without auxins. Supplement of 0.1 μM ABA helps inhibit precocious germination, especially root elongation.

4. The plantlets that are formed can be potted in soil and grown to maturity.

Exercise 6: Haploid plant production.

4.6.1. Anther and pollen culture

The cells of haploid plants, derived from pollen grains, contain a single complete set of chromosomes (n) and these plants are useful in plant-breeding programmes for the selection of desirable characteristics. Haploid plants are rarely produced under natural conditions. The phenotype of the plant is the summation in expression of single-copy genetic information which can either be expressed in a dominant, semi-dominant or recessive fashion. If a gene is recessive, the trait which it confers will only be expressed when in a homozygous recessive condition. Thus heterozygous dominant/recessive gene combinations will be expressed as a trait confered by the dominant rather then the recessive gene. The purpose of anther and pollen culture is to produce haploid plants by the induction of embryogenesis from haploid spores, either microspores or immature pollen grains. These haploid plants can be used to produce homozygous breeding stocks via artificial doubling using colchicine or the midvein technique. Sometimes spontaneous doubling of the haploid genomes of the plantlets being regenerated from microspores / anthers can occur in which case fertile dihaploid plants can be produced in a single step. These techniques involve the culture on defined media of immature anthers or pollen (microspores) just reaching the first mitosis following meiosis. These procedures and the pretreatments employed prior to culture are aimed to switch the typical gametophytic differentiation pathway (formation of two cells, the vegetative and generative cell, the latter of which would normally divide subsequently to yield the sperm nuclei which migrate down the pollen tube that is essentially organised by expression of genes in the vegetative cell component) to a pathway of development instead which involves the division of the microspore into a mass of embryogenic cells which then eventually yield plantlets which can either be haploid or dihaploid depending on whether or not spontaneous doubling takes place. Whichever the case, the resulting plants form the basis of homozygous parental lines of great value to plant breeders since homozygosity might take up to 5 years to attain in, for example, a cereal crop by means of repeated backcrossing and selfing (the so -

called pedigree method - viz. reviews by [1]Morrison and Evans, 1988 and [2]Foroughi-Wehr and Wenzel, 1990 for advantages and procedures involved in haploid production).

In the event that only haploid plants are produced, these plants can be treated with colchicine (0.6%) to bring about chromosome doubling and from the resulting mixture of haploids, diploids, aneuploids and polyploids, stable diploids can be selected following assessments of chromosome numbers in individual plants using various cytogenetic techniques (root tip squashes, microdensitometry, fluorimetry with 4,6-diamidino-2-phenylindole (DAPI) stain, chloroplast number per guard cell, etc) or comparisons of morphology - haploids tend to be more strappy plants while polyploids tend to be larger than their diploid counterparts. One advantage of microspore culture particularly in the cereals is that the discrete totipotent cells make ideal targets for transformation since any regenerating transformants would be expected to yield more total transformants than chimeric ones. However, it must be realised that production of haploid plants by isolated microspore culture is still restricted to certain genotypes since this procedure is heavily genotype-dependent and also that it is not the total population of pollen in any single anther which can be switched into embryogenic pathways of development. It is usually only 1-5% depending on species and genotype which has the capacity to regenerate embryos and plants. Furthermore, when plants are regenerated, a large number of these can be albino (white) due to the presence of lesions in the chloroplast/nuclear genomes induced either by the culture processes themselves or by a combination of genotype/culture factors, eg sugar type used in medium, and the type of pretreatment applied to flower buds prior to culture. Haploid plants can also be produced via the culture of unfertilised ovules.

Haploid cells possess 1n number of chromosomes
Diploid cells possess 2n number of chromosomes

---

[1]R. A. Morrison and D. A. Evans (1988). Haploid plant from tissue culture: New plant varieties in a shortened time frame. *BIO/TECHNOLOGY* VOL **6**: p684-689.

[2]B. Foroughi-Wehr and G. Wenzel (1990). Androgenetic haploid production. IAPTC News letter.

The chromosome complement of these haploids can be doubled using colchicine (usually 0.5%) to yield fertile homozygous diploids (doubled haploids - dihaploids).

Example: A section of chromosome may contain the following genes:

$$
\begin{array}{ll}
\text{A B C D E} & \text{- dominant genes} \\
\quad\text{or} & \\
\text{a b c d e} & \text{- recessive genes}
\end{array}
\quad\ldots(1)
$$

A section of heterozygous chromosome is composed of a mixture of recessive and dominant genes and as a result recessive genes coding for disease resistance (eg. d) may not be expressed if paired with a dominant gene allowing disease sensitivity (eg. D).

$$
\begin{array}{ll}
\text{A b C d E} & \text{Heterozygous chromosome pair} \\
\text{A B c D E} & \text{No disease resistance}
\end{array}
\quad\ldots(2)
$$

If the DNA of a haploid plant derived from pollen, is doubled, a homozygous chromosome pair is formed where there is no dominance of one gene over another (i.e. no hidden genes). From (2) it is possible to obtain two homozygous chromosome pairs (double haploid = dihaploid):

$$
\begin{array}{ll}
\text{A b C d E}\ldots(3) & \text{OR} \quad \text{A B c D E} \\
\text{A b C d E} & \qquad\quad \text{A B c D E}\ldots(4)
\end{array}
$$

If D codes for disease susceptibility and d for resistance to disease then (4) is not disease resistant but (3) is disease resistant. By screening large numbers of homozygous plants it is possible to select those which have disease resistance or some other commercially important trait.

Cultural considerations

* Pollen grains must be at the uninucleate stage of development for the best embryogenic responses. In the case of *Nicotiana tabacum*, floral buds with the petals (corolla) barely visible beyond the calyx will probably contain anthers at the appropriate stage of development.

* Anthers should be taken from flowers produced during the beginning of the plants' flowering period.

* Low-temperature pretreatment of anthers for 2-30 days at temperature of 3-10°C may stimulate embryogenesis. High temperature 35-40°C can also be effective in some plants, eg brassicas. These pretreatments are believed to synchronise microspore development by blocking metaphase of the first mitotic division so that when microspores/anthers are placed in culture, a greater number are capable of switching to embryogenic development.

* It is important to determine the chromosome number of the newly formed plantlets because there may be considerable variation in ploidy levels, depending on the developmental events that led to embryoid formation.

* There are two main techniques used for doubling the chromosome number of haploid plants.
    1) chemically induced doubling with colchicine
    2) regeneration by tissue culture methods

In this practical, anther culture of oil seed rape, ie. rapeseed (if plants flowering at correct stage) and broccoli will be performed and anther flotation culture in liquid media as well as isolated microspore culture of these species will be practised.

The methodology used is based on:
1. Keller, W.A. (1984). Anther culture of *Brassica*. In: *Cell Culture and Somatic Cell Genetics of Plants*. Ed. I.K.Vasil, Academic Press, Orlando, pp. 302-310.

2. Keller, W.A. & Armstrong, K.C. (1979). Stimulation of embryogenesis and haploid production in *Brassica campestris* anther cultures by elevated temperature treatments. *Theor. Appl. Genet.* **55**: 65-67.

3. Reinert, J. & Yeoman, M.M. (1982). *Plant Tissue Culture - a laboratory manual*. Springer-Verlag, Berlin. (a tobacco procedure)

Materials

* Plants of *Nicotiana tabacum*, broccoli or *Brassica napus* in the flower bud stage of development
* Two scalpels, fitted with narrow blades (No. 11)
* Two pairs of fine-nosed forceps
* 9cm Petri-dishes (10)
* 250ml flasks, each containing 200ml sterile ROW
* 100ml initiation medium (Medium 1 - Table 3)
* 250ml beaker (sterile) for 70% EtOH solution
* 100ml bottle (sterile) for 70% EtOH dip solution
* Dissection microscope

Table 3. Composition of media for culture of anthers

Constituents of culture medium	Final concentration (mg/l)
B5 medium	Full strength
Glutamine	800
Serine	100
myo-inositol	5000
Sucrose	60,000
pH 5.7 - 5.8	
Filter sterilise	

Procedures for anther culture

1. Prepare the culture Medium 1 for anther culture using B5 inorganic salt powder supplemented with 60 g/l sucrose, 5µM 2,4-D and 5µM NAA. Weigh out the other components individually and dissolve thoroughly. Adjust the pH (5.7-5.8) and add 0.3% phytagel, autoclave and pour twenty 5cm plates.

2. Select inflorescences bearing floral buds with petals between 17 and 21mm in length.

3. Excise the entire inflorescences and immediately place the cut basal ends in a flask of water. Wrap flowers in moist tissue paper and place into watertight bottle.

4. Surface sterilise in the usual manner and store one batch at 35°C for 48h (high temp. treatment) and another batch at 4°C for 48h (low temp. treatment).

5. Make all necessary flow cabinet, instrument and hardware arrangements before decontaminating the flower buds.

6. Remove and discard floral buds of the incorrect size and decontaminate the remaining buds in 70% EtOH for 30s. Rinse the buds in SDW and store in a sterile petri-dish until required.

7. Transfer the buds to a sterile polypropylene disk for excision of anthers. Cut off the tip of the flower petals with the scalpel and using the dissection instrument carefully remove the anthers without injuring anther walls. Transfer the anthers to the culture medium. Injured anther walls will form callus, which is not desirable.

8. Sterilise instruments regularly each batch of materials.

9. After the anthers are plated, seal petri-dishes or repli-dishes and place a loose-fitting transparent plastic bag over each stock of dishes.

Anther flotation and microspore culture

Anther flotation cultures of broccoli and tobacco will be established in replidishes and in so doing compare the effects of low and high temperature pretreatments of flower buds and the effects of different anther densities on the responses of anthers.

The two temperature conditions will be:
    I. High Temperature pretreatment (35° C for 48h)
    II. Low temperature pretreatment (4° C for 48h)

Anther densities:
    A. 2 anthers per 1 ml Medium 2
    B. 4 anthers per 1 ml Medium 2
    C. 8 anthers per 1 ml Medium 2

Medium 2 contains NLN salts (Table 4) supplemented with 5µM 2,4-D and 5µM NAA.

Table 4. The composition of NLN media

Components	Final conc. (mg/l)	Stock conc. (g/l)	Volume of stock for 1 litre medium
Macro salts KNO_3 $Ca(NO_3)_2.4H_2O$ $MgSO_4.7H_2O$ KH_2PO_4	125.0 500.0 125.0 125.0	1.25 5.00 1.25 1.25	100 ml (10 x stock)
Micro salts H_3BO_3 $MnSO_4.4H_2O$ $ZnSO_4.7H_2O$ $Na_2MoO_4.2H_2O$ $CuSO_4.5H_2O$ $CoCl_2.6H_2O$	6.20 22.30 8.60 0.25 0.025 0.025	3.10 11.15 4.30 0.125 0.012 0.012	Dilute 1:4 with ROW, then add 10 ml (500 x stock)
Iron FeEDTA Na salt	40.0		
Vitamins Glycine Inositol Nicotinic acid Thiamine HCl Pyridoxine HCl Folic acid Biotin Glutathione L-glutamine L-serine	2.00 100.00 5.00 0.50 0.50 0.50 0.05 30.0 800.0 100.0	0.20 10.00 0.50 0.05 0.05 0.05 0.005	10 ml (100 x stock)

Note: Add stocks or chemicals in order to 800ml ROW and allow each to dissolve completely before adding the next. Add **sucrose** (130g/l) and bring up to 1 litre, pH to 6.0 and filter sterilise.

Proceed as shown in the diagram below.

```
┌─────────────────────────────────────────┐
│   Unopened flower buds harvested        │
│         at appropriate stage            │
└─────────────────────────────────────────┘
                    │
┌─────────────────────────────────────────┐
│  Flower buds surface sterilised in 10%  │
│    bleach and washed twice in SDW       │
└─────────────────────────────────────────┘
                    │
┌─────────────────────────────────────────┐
│  Flower buds placed in plastic bags on  │
│   top of moistened sterile tissue paper │
└─────────────────────────────────────────┘
                    │
┌─────────────────────────────────────────┐
│ Bags placed in appropriate pretreatment │
│              conditions                 │
└─────────────────────────────────────────┘
                    │
┌─────────────────────────────────────────────────┐
│ Exact stage of flower bud development confirmed │
│  as assessed by acetocarmine or aceto-orceine   │
│          staining of anther squashes            │
└─────────────────────────────────────────────────┘
                    │
┌─────────────────────────────────────────┐
│ Buds immersed briefly in petridish of   │
│ 70% ethanol for 5-10s followed by a     │
│              wash in SDW                │
└─────────────────────────────────────────┘
                    │
┌─────────────────────────────────────────────────┐
│  Anthers dissected out from buds and placed     │
│  onto liquid Medium 2 at various densities      │
│       (A,B,C). Use plate set-up as overpage     │
└─────────────────────────────────────────────────┘
                    │
┌─────────────────────────────────────────────────────┐
│ 1ml aliqouts of sterile silver nitrate pipetted     │
│ into outer cells of replidishes to prevent          │
│ excessive ethylene build-up in cultures and         │
│ dessication of anthers during culture               │
└─────────────────────────────────────────────────────┘
```

Layout in replidishes

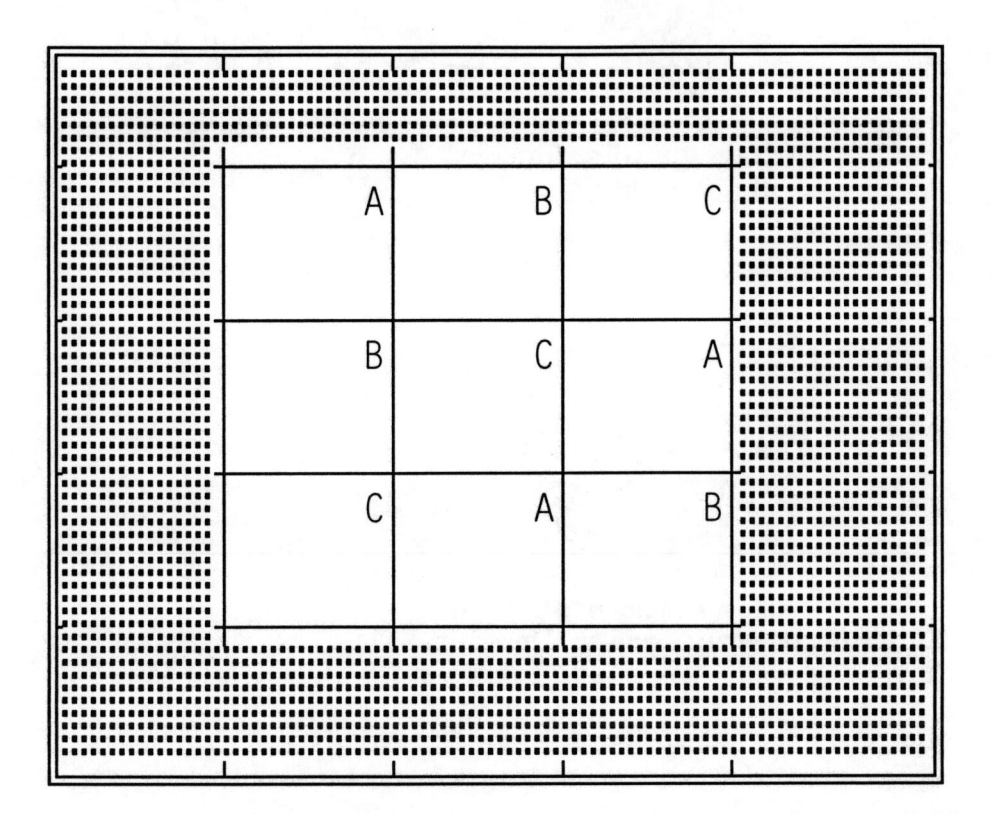

Treatments (either liquid or solid media) allocated to the inner 9 cells in a latin square arrangement, ie A,B,C; B,C,A; C,A,B.

A 20mg/l aqueous silver nitrate solution (SN) allocated to outer ring of 16 compartments as ethylene absorbant.

After setting up one replidish for each temperature and each species, seal the edges with nescofilm.

Incubate dishes at 25° C in the dark.

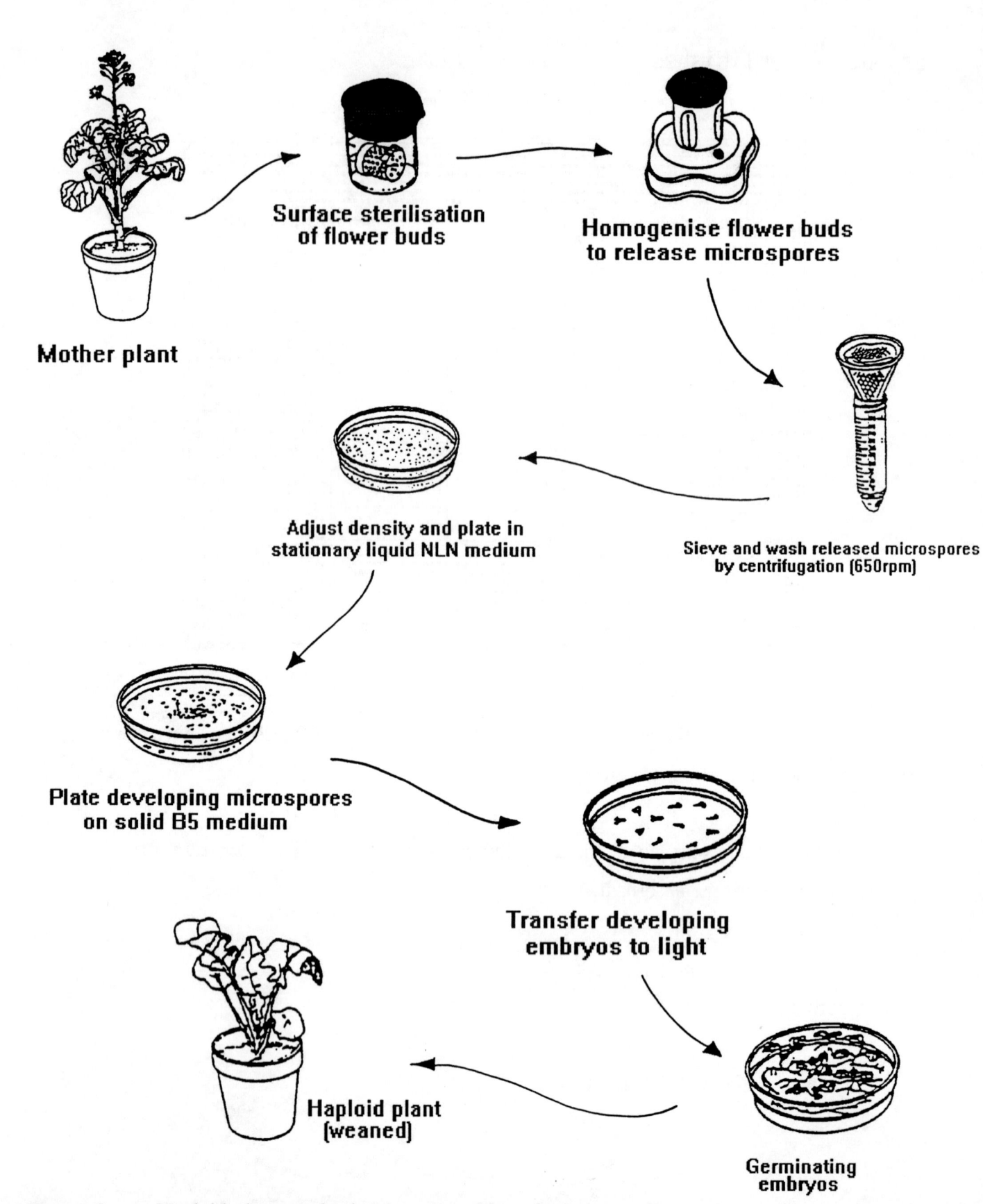

Figure 3. Haploid plant production in rapeseed by microspore culture

| Microspore Culture |

Follow the procedure outlined in the diagram on the next page. Medium NLN = Medium 2. Do not forget to prepare 300ml B5 salt medium with 6% sucrose for washing microspores following filtration.

General observations

* The first-emerged, easily separable plantlets are nearly always haploids; whereas plantlets emerging later may originate from non haploid callus tissues. Therefore, discard anthers after removing the first-emerged plantlets.

* Rooting occurs on the initiation medium (½ strength) without growth regulators solidified with 8 g/l agar.

* After root formation, collect root tips from each plant for cytological evaluation of ploidy level.

Reconstitution of diploids from haploids

* Several techniques can be used to obtain diploids from haploid plants. Usually it is desirable to retain the haploid plant for further study and a procedure is required that does not destroy nor chemically alter the haploid plant. A tissue culture technique works well for this purpose and makes use of the tendency of the leaves midvein to produce both haploid and diploid plantlets.

> **Exercise 7: *In vitro* seed culture**

4.7.1. Seed germination

Many plant species can be successfully introduced to *in vitro* culture making use of vegetative explants (e.g. nodes, leaf pieces, stem segments, anthers, ovules, apical tips etc.). However, if the use of these explants leads to the destruction of a rare plant, if the vegetative explants are difficult to sterilise or the explant is not appropriate for the envisaged experiments, then seeds can sometimes be used as a source of disease free explants. The embryos can be dissected out of developing (embryo rescue) or mature seeds (embryo culture), or the seed can be allowed to germinate *in vitro* thus producing a sterile source of young nodal and apical tip tissues for micropropagation. Shoots obtained from aseptically cultured seeds are also normally free (not always) of systemic diseases. The use of seed for micropropagation can be useful for homozygous plant species (e.g. wheat or dihaploid tobacco) but for heterozygous plant species (e.g. grapevine) there is no use for this technique unless part of a plant breeding project.

Uses of *in vitro* seed germination:

1) To increase the germination rate of rare or difficult-to-germinate seeds.

2) To obtain sterile material of plant species whose parent vegetative parts are normally difficult to sterilise.

3) To obtain disease-free plant material

Procedures for seed sterilisation

A) **Pre-sterilisation treatments**
   1. Hot water treatment: 50°C for 10 min, for dirty seeds 100°C for 1 min, for seeds with hard testa (e.g. *Acacia*)

   2. Soaking in running tap water for 1-2 h

   3. Washing in 70% EtOH (v/v) for 30s to 2 min with certain fruits (cycads)

   4. Up to three rinses in sterile distilled water

   5. Seeds within an intact fleshy coat (testa) are normally sterile and only the seed coat need be sterilised by submerging for 20-30 min in 10% v/v bleach solution after which the seed is excised.

B) **Sterilisation**
   1. 20 - 30 min in 10 - 50% (v/v) bleach, possibly under negative pressure, followed by three rinses in SDW.

C) **Post sterilisation**
   1. Wash three times in sterile water and either germinate in SDW, damp sterile filter paper or on diluted nutrient medium (e.g. Murashige & Skoog, 1962).

4.7.2. Embryo rescue.

The aim is to dissect out the immature or mature embryo from a seeds and to culture the embryo in *in vitro* conditions.

Uses:

1. **Plant breeding**

    Most species will not cross successfully with another species to produce fertile plants. Even if pollination and fertilisation occur between two different species, the embryo will often not develop due to inhibition of the surrounding endosperm tissue. If the young immature embryo is 'rescued' and placed on an appropriate medium, then the embryo may develop into a new plant (sucrose and glutamine are important requirements for the growth of an immature embryo). This embryo can also be induced to form somatic embryos (bipolar embryos) directly by subjecting the excised embryo to high 2,4-D for a short period and then placing on a medium without growth regulators (called somatic embryogenesis).

2. **To overcome dormancy**

    e.g. Tree paeony - epicotyl dormancy. The seed usually takes 2 years to germinate because the epicotyl requires 2 years of winter chilling in order to grow. The germination can be speeded up in the following way.
    - sterilise seed and dissect embryos
    - place embryo on MS medium
    - 5 weeks at 20°C - root and cotyledons grow
    - 5 weeks at 4°C - to overcome epicotyl dormancy
    - transfer to 20°C for new shoot growth

3. A lot of work is carried out on forestry trees e.g. conifers using embryos (immature and mature) because they are more regenerative than older tissue.

General points:

A high cytokinin:auxin ratio tends to give shoots. A low cytokinin:auxin ratio tends to give roots. Cytokinins are used in the range: 2.5 - 25µM depending on plant species and morphogenic response required. Auxins are used in the range: 0.5 - 25µM depending on plant species and morphogenic response required.

Growth responses

Cytokinins	Auxins
BAP shoot production 2iP and development kinetin	IBA rooting NAA rooting and/or callus production 2,4-D callus and/or somatic embryogenesis

Exercise 8: Microdissection.

4.8.1. Micrografting.

One of the major problems associated with the clonal propagation of ornamental or forestry trees of proven quality lies in the fact that mature tissues generally lack the ability to regenerate roots or produce orthotropic growth habits ([3]Hartmann, Kester and Davies Jr, 1990). This is particularly true of conifers, fruit and nut trees, and the viable method of cloning available is the conventional propagation of mature trees through grafting, using terminal shoots as scions.

Recently, much attention has been focussed on micropropagation and tissue culture techniques, offering not only the unlimited production of selected clones but also the potential of advances in breeding via polyploidy, haploid and resistance breeding. Amongst species which do respond to micropropagation techniques, best results have been obtained using explants from either juvenile tissue or from epicormic shoots from the proximal parts of the tree still exhibiting juvenile characteristics. Materials can also be rejuvenated via serial micrografting of mature scions onto seedling root stocks.

Micrografting has been also used in fruit and nut tree improvement programmes for the production of disease-free scions and for the study of the histological nature of graft unions. The aim of this study was to get familiarised with micrografting techniques.

---

[3]Hartmann, H.T. and Kester, D.E. (1990). *Plant propagation: Principles and practices 4th Ed.* Prentice Hall International Inc, New Jersey, USA.

Materials

* Plain MS liquid medium in 4" tubes, each containing 2 ml medium.
* Seedlings for use as rootstock, eg. citrus, coffee, etc, germinated on Milcap polyester plugs.
* Either *in vitro* cultures or glass house shoot tips to be used as scions.
* Dissecting instruments, including a binocular dissecting microscope.

Procedure

1. Dissect carefully the scion from the shoot tips of either *in vitro* shoot cultures or glass house shoot cuttings in a drop of sterile ROW to prevent dessication of the scion.

2. Decapitate the seedling to be used as rootstock, and make a longitudinal incision through the axil itself to a depth of about the size of the scion.

3. Now, carefully insert the scion into the incirsion and place the rootstock back into a fresh tube containing MS liquid medium prepared earlier.

4. Seal tubes and incubate at 25°C, 16h photoperiod.

5. Monitor the growth and development of the graft union and if necessary, make histological sections to study the development in detail.

Exercise 9: Protoplast culture techniques.

Introduction

Plant protoplasts are "naked" plant cells. They are cells from which the cell wall has been removed by either mechanical or enzymatic means leaving the outer plasma membrane fully exposed. This is the only barrier between the external environment and the interior of each protoplast. The potential uses of plant protoplasts are:

1) **Protoplast fusion.** The fusion product can be carefully nurtured to produce a hybrid plant of two incompatible plant species (inter- or intra-specific hybrids). One of the uses of this protoplast fusion is to transfer disease resistance from one plant to another.

2) Isolated protoplasts are able, under certain conditions, to take up DNA which may be inserted into the plants genome and subsequently expressed in a viable plantlet (direct DNA uptake, one route of genetic transformation).

3) The cultured protoplast rapidly regenerates a new cell wall, and this developmental process offers opportunities to study wall biosynthesis and deposition.

4) Populations of protoplasts can be studied as a single cellular system and microbiological methods have been developed for the selection of mutant cell lines and the cloning of cell populations.

> **General information**

The chief function of the cell wall (which is permeable to most organic and inorganic molecules) is to exert a wall pressure on the enclosed protoplast and thus prevent excessive water uptake leading to bursting of the cell plasma membrane. Before the cell wall can be removed, the cell must be bathed in an isotonic plasmolyticum, which is carefully regulated in relation to the cells osmotic potential. The sugar alcohols sorbitol or mannitol are used to provide a suitable osmotic pressure outside the cells in order to prevent the naked cell from bursting due to excessive uptake of water. In addition these compounds are not metabolised by the plant cells in contrast to sucrose which has a morphogenic effect.

Enzymatic procedures are used to remove the plants cell wall. The enzymes used are extracted from fungi which parasitise plant cells. Macerozyme digests pectin and cellulase digests cellulose.

4.9.1. Protoplast isolation and culture

> **Basic procedure**

1. Surface sterilisation of leaf samples (if greenhouse plants are used). This stage can be avoided by the use of *in vitro* plantlets.

2. Peeling off the lower epidermis or slicing the leaf tissue to increase surface area which facilitates enzyme reaction.

3. Mixed enzyme treatment of leaf tissue and incubate in dark at 25°C for a specific length of time.

4. Purification of the isolated protoplasts by removal of enzymes and cellular debris by centrifugation.

5. Transfer of known density of the protoplasts to a suitable culture medium to encourage cell wall formation, cell division and callus formation.

> **NB:** * To maintain membrane stability calcium ions (Ca^{2+}) are included in the medium.
> * The purification step is important because it is vital to remove all traces of enzyme from around the protoplast which are easily damaged by their action.

Materials

* Collect mature leaves (*ca.* 25cm long) from greenhouse grown plant which have not yet begun to flower or use the leaves from young *in vitro* cultured plantlets
* Equipment and chemicals for leaf sterilisation
* Two scalpels fitted with new, sharp blades, or use steribead razor blades
* Two pairs of forceps
* 50 ml protoplast isolation medium (Table 5) - filter sterilised
* 200 ml protoplast wash and culture medium (Table 6) - autoclaved
* 200 ml protoplast purification medium (Table 7) - autoclaved
* 2 x 45-µm pore-size nylon or stainless steel mesh sieves
* Sterile Millipore filter holder or similar ultra-filter equipped with a membrane filter, 0.22-µm pore-size
* 2 x 9cm sterile Petri-dishes
* 2 ml pre-sterilised wide-bore pipettes
* 10 ml pre-sterilised wide-bore pipettes
* Pre-sterilised plastic centrifuge tubes
* Test tube holder to fit centrifuge tubes

> **Tobacco protoplast isolation medium (TPIM):**
>
> Full MS medium + 10% mannitol
> + 0.5% Cellulose (1 or 2% for rapid release)
> + 0.1% Macerozyme (0.5% for rapid release)
> + 2% PVP
> + 0.15% MES buffer
> pH 5.8
>
> Note: **DO NOT HEAT** (heating will denature the enzyme) **SO FILTER STERILISE**

> **Tobacco protoplast wash and culture medium (TPWCM):**
> Full MS medium + 10% mannitol
> + 15µM 2,4-D
> + 5µM BAP
> + 1% glucose
> + 2% sucrose
> pH 5.8
>
> * The purification medium remains the same as given in Table 7.

Procedure for isolating protoplasts from leaf tissue of tobacco.

1. Transfer *in vitro* or glasshouse grown tobacco plants to dark culture 48 hours prior to taking leaves in order to reduce the starch levels in the cell (avoids cell bursting during isolation).

2. Select young leaves from plants growing in the greenhouse or young expanding leaves from vigorous *in vitro* plantlets.

3. Surface sterilise glasshouse grown plantlets in the usual way. Peel the lower epidermis and place the peeled side of the leaf down onto the enzyme solution (10ml TPIM for each petri-dish). Cover the surface area of the dish with peeled leaf material.

4. *In vitro* derived tissues do not need to be sterilised and have the advantage of being juvenile. In this case the leaves are cut up into very fine strips with the aid of a razor blade or fresh scalpel blade and placed into the enzyme solution.

5. Incubate at 25°C in the dark for 9-15h (varies depending on the plant species and the concentration of the enzyme) on a eccentric arm shaker (<20rpm). During incubation most of the leaf material is broken down into a cell/protoplast mixture. The suspension should be checked periodically to determine when the majority of protoplasts has been released into suspension. Prolonged culture can destroy the protoplasts (over digestion).

6. Using a sterile 10ml wide-bore pipette transfer the protoplast mixture to a sterile centrifuge tube through a sterile gauze filter (previously moistened with few ml TPWCM). The filter removes any undigested leaf material. Wash the filter through with few ml TPWCM to wash through all released protoplasts attached to the debris collected in the filter. Centrifuge the suspension at 650rpm for 10min.

Purification of protoplast suspension

1. Pipette off the supernatant leaving the protoplast pellet at the bottom of the tube. To obtain floating epidermal protoplasts suck off top 3ml of supernatant and purify by floating on a 0.45M (22%) sucrose (filter sterilise) solution. Centrifuge for 10 min at 650rpm. Bouyant, viable, intact protoplasts collect as a green ring in the high sucrose medium. Carefully remove this ring using a sterile pasture pipette and collect in a fresh centrifuge tube.

2. Resuspend gently with approximately 5ml TPWCM and centrifuge for 5 min at 650 rpm; then pipette off the supernatant, leaving the protoplast pellet at the bottom of the tube. Add 5 ml fresh TPWCM and resuspend gently.

3. Repeat the above washing procedure twice.

4. Following the third wash, resuspend the protoplasts in 2ml TPWCM. Determine the density of purified protoplasts using a drop of protoplast suspension on a counting slide under the microscope (see Growth and Development assessments) and adjust the density to approximately 1×10^5 protoplasts/ml by the addition of culture medium. Check viability by adding a drop of 0.01% Fluoreacein diacetate (FDA) to a drop of protoplast suspension and viewing under uv light. Viable protoplasts will fluoresce brightly.

5. Pipette single drops of the protoplast suspension in culture medium onto a 5cm sterile petri-dish (10 drops/dish). Close the petri dish and seal with Nescofilm. Incubate in a humid chamber (eg. sandwich box with 10-15ml sterilised ROW at the bottom or in a square replidish in which the outer wells contain 1-2ml sterile ROW and the inner nine wells contain the protoplasts).

6. After a few days, viable protoplasts should regenerate new cell walls. Cell walls can be detected by staining the cells with Calcofluor White. This white dye binds to wall material and exhibits fluorescence on irradiation with blue light or UV light. Once the cell wall has been formed, each new daughter cell cluster should start to divide to form a microcallus.

7. Transfer microcalluses to a small amount of fresh culture medium with 4 - 6% mannitol (lower osmotic pressure) either in liquid or solid form and incubate for several weeks to allow further callus proliferation.

8. Large calluses can be picked out and placed onto shoot regeneration medium if regeneration does not occur spontaneously on the normal culture medium.

Procedure for isolating protoplasts from yam cell suspension cultures.

1. Transfer cell suspension (3-5 days after subculture) to 10 ml centrifuge tubes and centrifuge at 650rpm for 5 min.

2. Resuspend cells in protoplast isolation medium (TPIM). Transfer the solution to 5 ml petri dishes (10 ml per dish) and incubate on an oscillating in the dark at 25°C overnight.

3. Gently break cell aggregates by pipetting the cell suspension through a sterile Pasteur pipette every 2h and monitor protoplasts release under an inverted light microscope. In most cases, protoplast release from cultured cells is generally completed within 6 to 7h.

Table 5: Protoplast isolation medium (YPIM)

Component	Molecular Weight	mM	g/100ml
$CaCl_2.2H_2O$	170.48	9.00	0.15
$NaH_2PO_4.2H_2O$	156.01	0.90	0.015
Mannitol (10%)	182.17	550.0	10.0
MES buffer	195.2	7.70	0.15
PVP (2%)	44,000	0.45	2.0
pH 5.8	Autoclave		
Cellulase (0.5%)			0.5
Macerozyme (0.1%)			0.1
Filter sterilise			

4. Transfer the digest to 10 ml centrifuge tube through a sterile stainless steel sieve (*ca.* 50μm) and collect the protoplasts by centrifugation at 650rpm for 5 min.

5. Decant the isolation medium and wash protoplasts at least three times in protoplast wash and culture medium (TPWCM).

Table 6: Protoplast wash and culture medium (TPWCM)

Component	Molecular weight	mM	g/100ml
MS medium			0.471
$CaCl_2.2H_2O$	170.48	9.00	0.15
$Na_2H_2PO_4.2H_2O$	156.01	0.90	0.015
Mannitol (10%)	182.17	550.0	10.0
2,4-D	221.0	0.14	0.003
BAP	225.2	0.04	0.001
Glucose (1%)	180.16	55.51	1.0
Sucrose (2%)	342.30	58.43	2.0
pH 5.8	Autoclave		

6. Resuspend protoplasts in 1 ml YPWCM, pool samples and assess their viability using FDA. Performed in a similar way to tobacco protoplasts as described earlier.

7. If the viability is low or in the case where cell debris is present in excessive amounts, the protoplasts are purified by flotation on a sucrose medium - protoplast purification medium (YPPM).

8. Release the resuspended protoplasts (step 6) gently using a pasteur pipette to float onto YPPM contained in a fresh centrifuge tube and centrifuge for 5 min at 650rpm.

9. Collect and wash the bouyant, viable, intact protoplast ring formed on the YPPM to remove the purification medium and assess the density as well as their viability once again before culturing.

10. The protoplast yield is determined by placing a few drops of protoplasts onto a counting slide and recording the numbers of protoplasts present in 10 separate fields of the microscope at a given magnification.

11. The number of protoplasts expected to be present in each ml of the protoplasts suspension is calculated using the formula $V=\pi r^2 \times 0.8$ (where V= volume, r=radius of the field of microscope) and converting the volume to 1 ml and finally to the total measured volume of protoplasts suspension (NB. there are 1000 mm^3 in 1cm^3)

12. Culture the protoplasts in a similar manner to tobacco protoplasts (c. 10^4 protoplast per ml)

Table 7: Protoplast purification medium (YPPM)

Component	Molecular weight	mM	g/100ml
MS medium			0.471
Sucrose (22%)	342.30	642.71	22.0
2,4-D	221.0	0.14	0.003
BAP	225.2	0.04	0.001
pH 5.8	Autoclave/filter sterilisation		

Background information on protoplast isolation

* Enzymes - these are extracted from saprophytic fungi which digest plant material in the soil environment. Fungi such as *Trichoderma* and *Penicillium* fall into this group.

Macerozyme - digests pectin which cements cells together.
Cellulase - digests cellulose contained in cell wall.

> **Osmoticum:** The cell wall exerts a cell wall pressure which acts against the osmotic pressure of the cytoplasm within the cell. The cell wall prevents the cell from bursting under unfavourable conditions (eg. fluctuating osmotic environments)

A protoplast has no cell wall to prevent bursting. To avoid such bursting or shrinkage, the osmotic potential of the surrounding culture medium needs to equal or slightly exceed the cell osmotic pressure (through the use of isotonic or hypertonic solutions). Mannitol at a concentration of 10% gives similar osmotic pressures to those of the cell sap of most plants. Mannitol is not metabolised by plant cells. During purification, a 22% sucrose gradient (equivalent in osmotic potential to 10% mannitol) is used to float and separate the protoplasts from other cell debris. This is because sucrose is much denser than mannitol and creates a differential density gradient during centrifugation. Until the protoplasts regenerate cell walls, mannitol or sorbitol is required in the medium to keep the osmotic pressure high. Over a period of 2 - 3 weeks the osmoticum concentration can be reduced and eventually omitted altogether. If the osmoticum is kept high for too long a period, callus formation slows down due to an inhibition of cell division induced by physiological stress (the raised osmotica cause lower water contents in the cytoplasts of cells amongst other things).

> * **Membrane stability**
> Ca^{2+} ions are included in the protoplast medium to stabilise the cell membrane.

4.9.2. Growth and development assessment.

(A) Cell density

Place a drop of cell suspension/protoplast suspension onto a counting slide made of 0.8mm deep walls, constructed from pieces of standard glass slide glued with epoxy resin to a complete slide. Cover the sample with a clean cover slip and count the number of cells/protoplasts present in 5 to 10 separate fields of the microscope at a given magnification and get an average. Using a graticule slide calculate the diameter of each field.

Having determined the radius of a field, the volume of suspension examined under each field can be calculated as follows:

$$V = \pi r^2 \times 0.8 \quad \ldots \ldots \ldots 1$$

Where: V = volume and r = radius of field

Then convert this volume to 1ml (1cm^3). The number of cells present in each ml of culture can be calculated. If the cell/protoplast density is too high then it can be diluted using the relevant culture medium, but do not forget to account for this dilution in your calculations.

(B) Viability

There are various dyes available for viability assessments. These work on the basis of one of two principles, namely:

Exclusion
* Viable membranes exclude coloured dyes like analine blue (Evans Blue) or phenosafranine red. Cells which stain up in such dyes are non-viable.

> ***Conversion of dyes to easily identified products***
> * Viable cells contain esterase enzymes capable of converting substrates to easily recognisable products. Fluorescein diacetate (FDA) dissolved in acetone is converted to fluorescein on membranes and in the cytoplasm of viable cells. The fluorescein produced (and fluoresces brightly under B filter uv) is retained within vacuoles and the cytoplasm of viable cells because of the presence of intact membranes.

> ***Other dyes which can be used:***
> * Tetrazolium dyes are reduced to formazans by functional lipases and dehydrogenases. These accumulate in viable cells as strongly coloured granules that can be easily recognised in visible light.

Procedures

Exclusion dyes.

A drop of 0.1% aqueous analine blue is added to a 5ml sample of cell suspension. Leave for 15 min before a few drops are examined with visible light. Determine the percentage of cells which DO NOT stain up strongly with the blue dyes. These are the viable cells.

Enzymic methods

Make up a stock solution of FDA (5mg/ml acetone). Place one drop of this stock into 10ml SDW for cell suspension (or culture medium for protoplasts so as to make a final concentration of c. 0.01%). Add this solution in a 1:1 ratio to the cell/protoplast suspension and allow to settle for 2 min. Place a few drops of treated suspension on a slide and examine under the IMT-2 inverted microscope using UV light. **WARNING:** UV light can seriously damage your eyes if the correct filter is not used. Please ask if uncertain about which filter is to be used on the IMT-2 microscope. Do also remember that FDA loses its fluorescence properties after about 2h in daylight and so any diluted FDA solutions have to be prepared fresh each time using the acetone stock solution which can be kept safely in a screw capped vial in a freezer at -20°C.

Calculate the % cells which show strong, positive bright green fluorescence due to the presence of fluorescein. This represents % viability of the cell population.

Calcofluor white:
Shows up the presence of cell walls, when UV light is passed through the suspension. Formation of the cell wall is an indication of the regeneration of protoplasts into callus.

4.9.3. Protoplast fusion

Several approaches to altering the genetic content of somatic plant cells are available and one of the most significant of these is the fusion of protoplasts within species, between species, between monocots and dicots, and between plants and animals. Cells containing transferred organelles are called cytoplasmic hybrids, cybrids or heteroplasts. Those containing nonidentical nuclei are referred to as heterokaryons or heterokaryocytes; when these nuclei fuse, a hybrid cell is formed. There are two main ways of inducing fusion between protoplasts: electrofusion and chemical fusion.

(A) Electrofusion

Before electrofusion can take place, the protoplast suspension needs to be purified on a density gradient so that it is free from cell and membrane debris and also washed several times to remove the high electrolyte contents of the protoplast media used during the isolation, purification and wash stages of protoplast preparation. The presence of contaminating debris and ions adversely affects the electrofusion process. Electrical current should pass straight through touching protoplasts without being directed through any ionically charged substances in the imbibing medium. The removal of debris is accomplished by sucrose flotation and the removal of mineral ions by repeated centrifugation / resuspension (2-3x) in 0.5 - 0.6M mannitol.

> ***The principle of electrofusion***
>
> An alternating current (AC) is passed through the protoplast suspension so that lining up of the protoplasts occurs, a pearl chain formation. A short pulse ("Zap") of direct current (DC) is then applied which causes dissociation of the protoplast membrane. Where two protoplasts are touching each other, the membrane dissociates at the point of contact, enabling the two protoplasts to fuse as the dissociated membranes coalesce together.

Procedures

* *Preparation of protoplasts and equipment*

1. Remove cell debris from protoplasts using sucrose gradient (carried out as under the "protoplast isolation and purification" procedure described earlier in 4.9.1).

2. Suspend and wash protoplasts in 10% mannitol twice to remove any ions which may interfere with the fusion procedure. Store protoplasts on ice until used.

3. Dispense 2ml protoplast from each clone/plant species into a centrifuge tube. Adjust density to 1×10^5 cells/ml.

4. Using sterile pasteur pipette dispense a layer of protoplasts into a sterile fusion chamber. Ensure that the entire area between the electrodes is filled with liquid.

5. Wrap dishes to maintain sterility. Leave electrode contacts exposed.

Preparation of fusion equipment

1. Fix petri-dish to microscope stage with clips and connect wire to the electrodes.

2. Set timer for 1 min and switch on the oscilloscope power supply for electrodes.

3. Switch on function generator - 500 kHz AC; 6 volts; turn the Amp switch off.

4. Switch on DC power generator - supplies up to 400 volt pulses maximum. **Note**: voltage set at 200V for sinking protoplasts but lower for floating protoplasts.

5. Turn the amp switch on now, and apply the AC voltage for 1 min to line up the protoplasts (observe the pearl chain formation).

6. After 1 min give 2 pulses of DC voltage (varies accordingly with plant species to induce fusion.

7. Immediately turn down or switch off the amp switch off AC voltage supply.

8. Observe those protoplasts which have undergone fusion. Leave to stabilise for 30 min.

9. Plate protoplasts out onto nutrient medium (TPWCM or YPWCM).

(B) Chemical fusion

This technique makes use of polyethylene glycol (PEG) to induce the fusion of protoplasts. The fusion medium is given in Table 7.

Table 8. PEG protoplast fusion medium (YPFM)

Component	Quantity
PEG 6000	4.5g
$CaCl_2$	15.5mg
KH_2PO_4	0.9mg
Sterile Distilled water	10ml
pH 6.2	

Place a drop of purified protoplasts suspended in 0.6M (10%) mannitol in the centre of a sterile microscope slide. Carefully pipette four drops of PEG solution around the protoplast suspension drop. After 15 min look for signs of fusion under the microscope. Any fused protoplasts can be carefully pipetted out and placed into culture medium.

Exercise 10: Solving problems in micropropagation.

4.10.1. Vitrification.

Vitrification (or hyperhydricity) of *in vitro* cultures is typified by the glassy appearance and brittleness of leaves and stems. Vitrification occurs as a response to high humidity or watery conditions e.g. in a liquid medium. Vitrified tissue has poorly developed lignin and cellulose in the cell walls. Osmosis: diffusion of water across semi-permeable membrane from an area of low osmotic potential to an area of high osmotic potential.

A cell contains solutes which give it a certain osmotic potential (pressure). A high osmotic potential will encourage diffusion of water into the cell. The cell membrane is a semi-permeable membrane). However, the cell wall will also have a restraining effect so that the cell can only take in a certain amount of water until it is completely turgid. When the cell is turgid, the pressure of the wall (turgor pressure) restricts any more water diffusing into the cell even if there is a difference in osmotic pressure inside and outside the cell.

In vitreous cells, the reduced lignin and cellulose mean that the cell wall turgor pressure is less and allows more water to diffuse into the cell. So the cell becomes hypertrophic - enlarged and 'watery'.

In vitrified epidermal tissues it has been found that the stomata lack functional guard cells; they do not close in response to low humidity - again the fault of the cell wall. In fact, the lack of functional guard cells and cuticular wax is common in many tissue cultured plants since plants in general will often modify their internal structure as an adaptation to humidity.

Obviously some plants are more susceptible to the high humidity enclosed environment than others, eg. carnation - is commonly cited as having problems with vitrification. It is not only the plant, but also the culture conditions such as: (i) pH of medium, (ii) length of time in culture and (iii) mineral salts and hormones.

How to reduce vitrification:

> **1. Reduce humidity of culture vessel**

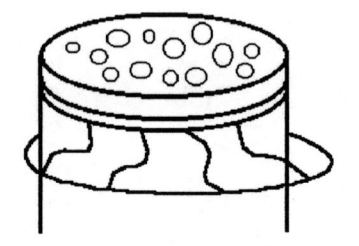

Condensation on polypropylene covers can block pores so that gaseous exchange cannot occur as readily and this results in a build up of H_2O vapour, ethylene and CO_2 - all may effect the morphogenesis of the plant. H_2O and ethylene in particular can induce vitrification in a sensitive plant.

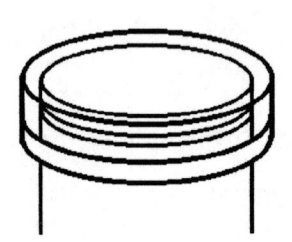

Polypropylene caps
- allow more movement of gaseous exchange.

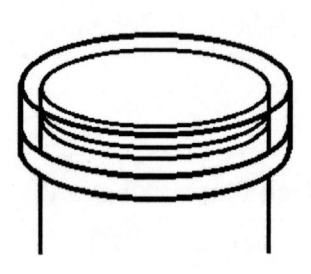

Jars with loose fitting lids - induce normal growth from vitrified tree paeony; but increase chances of contamination. In this case it is very important to keep the vessels and growth room clean and dust free.

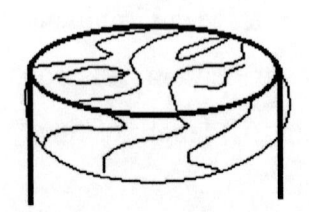

Cling film - allows good exchange of gases and water vapour, thereby reducing vitrification, but care has to be taken to avoid the culture and medium drying out.

2. Agar - type and concentration

A high agar concentration will decrease the amount of water available to the plant tissue - since the more agar molecules will bind with more water molecules. This often reduces vitrification but also reduces the availability of other substances such as hormones and mineral salts, which may in turn reduce the growth and/or multiplication rate.

3. Effects of NH_4^+ and Ca^{2+}

Quite a few papers have cited that reducing the NH_4NO_3 in the medium (or even in some cases omitting it) can reduce the effect of vitrification e.g. chestnut and willow. Several suggestions have been made for why the NH_4 ion has an effect on vitrification.

Deficiency of lignin and cellulose is known to result from a fall in C/N ratio after an increased uptake of N from the NH_4/NO_3 ratio. NH_4^+ ions are assimilated faster than NO_3^- ions. [4]Vieitez *et al.* (1985) speculates that this rapid uptake may increase the consumption of carbohydrates sufficiently to divert them from the metabolic pathways leading to the synthesis of lignin. An increase in vitrification, caused by an increase in NH_4^+ may be mediated by an increase in glutamate dehydrogenase activity.

Also a higher lignin level of normal tissue were seen to correspond to a higher phenylalanine ammonia lyase activity in carnation ([5]Kevers and Gaspar, 1985) so the more the ammonium compounds are metabolised, more lignin is synthesised. A medium rich in ammonium may become saturated and reduce overall lignin production in tissue leading to vitrification.

---

[4]Vieitez, A.M., Ballester, A., San-Jose, M.C. and Vieitez, E. (1985). Anatomical and chemical studies of vitrified shoots of chestnut regenerated *in vitro*. *Physiologia Plantarum* **65**, 177-84.

[5]Kevers, C. and Gaspar, T. (1985). Vitrification of carnation *in vitro*: Changes in ethylene production, ACC level and capacity to convert ACC to ethylene. *Plant Cell Tissue Organ Culture* **4**:215-223.

Ca^{2+} ions have also been reported as important determinants of vitrification *in vitro*: a high Ca^{2+} in the medium tends to reduce vitrification.

4. Effect of medium pH

The medium pH is important since it affects the availability of ions in the medium to the plant, e.g. Schwartz and Fleminger (1987) found that vitreous leaf development was greatly reduced in apices cultured in a medium buffered with 10mM MES to pH 5.2 - 5.4 as compared to an unbuffered drop to pH 4. In addition, the pH of the medium affects the viscosity of the agar. At pH 6.0, 8% agar gives a firm substrate whereas at pH 5.0, 8% agar is softer.

5. Effects of growth substances

Growth substances such as auxins and cytokinins may affect plants via induction of ethylene which may build up in an enclosed vessel, especially in ageing cultures where the frequency of senescent leaves increases. It is often found that high cytokinin tends to induce vitrification in ethylene sensitive plants, although Lesham and Sachs (1985) have found that high NAA increased abnormal shoot tips whereas high BAP reduced abnormal shoot tip development in carnation. So it is not a straightforward problem and there are no clear cut solutions. These factors may be interrelated and it is difficult at this stage to isolate the causes. Several biochemical pathways may be affected by a high mineral/high hormone/high humid environment resulting in reduced lignification of cell walls and increased hypertrophy of tissues.

Vitrification can be beneficial - Lesham wrote a paper in 1986 on somaclonal variation generated from vitrified carnation, where abnormal (vitrified) plants gave rise to normal plants. But it is often detrimental and there is a need to adapt both the medium and culture conditions to overcome it.

The following experiment aims to investigate the culture and medium conditions required to reduce vitrification in shoot cultures of a number of herbaceous and woody plants.

Treatments (including 30g/l sucrose)

A	MS + 4 g/l	agar, no growth regulators, polypropylene covers
B	MS + 4 g/l	agar, no growth regulators, cling film
C	MS + 4 g/l	agar, no growth regulators, aluminium foil
D	MS + 8 g/l	agar, no growth regulators, polypropylene covers
E	MS + 8 g/l	agar, no growth regulators, cling film
F	MS + 8 g/l	agar, no growth regulators, aluminium foil
G	MS + 4 g/l	agar, 10 mg/l cytomix, polypropylene covers
H	MS + 4 g/l	agar, 10 mg/l cytomix, cling film
I	MS + 4 g/l	agar, 10 mg/l cytomix, aluminium foil
J	MS + 8 g/l	agar, 10 mg/l cytomix, polypropylene covers
K	MS + 8 g/l	agar, 10 mg/l cytomix, cling film
L	MS + 8 g/l	agar, 10 mg/l cytomix aluminium foil

Replications: 5 tubes per treatment

Materials

* Plant material e.g. carnation, coleus and apple (*in vitro* cultures)
* One roll of cling film
* One box polypropylene squares
* Some sterile aluminum foil squares
* One box polypropylene disks
* 60 tubes of the various MS medium (A - L) prepared as above
* Dissection instruments

> **Procedures**

1. Remove shoot clusters from tubes and dissect into individual nodal cuttings. Place onto the specific media in the tubes and cover with either polypropylene or cling film covers or aluminium foil.

2. Incubate at 19°C in a 16h light culture room.

3. Subculture every 4 weeks. Make notes of the physical appearances of microplants at regular intervals (every two weeks) and interpret the results in the light of the observations made with respect to optimising culture and medium conditions necessary for normal vigorous growth of each of the respective species investigated.

4.10.2. Browning

Establishment of *in vitro* cultures of several plant species, especially woody plants, is greatly hampered by the lethal browning of the explant and culture medium. Browning is considered genereally to result from the oxidation of phenolic compounds. These are released from cut ends of the explants, by polyphenoloxidases, peroxidases or air (oxygen). The oxidised products, like quinones, are known to be highly reactive and inhibit enzyme activity leading to the death of explants ([6]Hu and Wang, 1983).

The strategies employed to overcome or reduce the harmful effects of browning attempt to either neutralise or avoid the build up of toxic substances in the medium surrounding the explant in particular. These include:

i) choice of juvenile explant materials, or new growth flushes during the active growth period

---

[6]Hu, C.Y. and Wang, P.J. (1983). In: Handbook of plant cell culture, vol 1 (eds. Evans, D.A., Sharp, W.R, Ammirato, P.V., Yamada, Y.). Macmillan, New York, pp177-217.

ii) culture in darkness, especially in earlier stages of initiation
iii) transfer to fresh medium at short intervals
iv) culture in liquid medium
v) inclusion of antioxidants (antibrowning agents) in the culture medium, or soaking explants in water or solutions containing antioxidants prior to inoculation
vi) use of adsorbing agents, such as activated charcoal, polyvinylpyrrolidone (PVP)
vii) choice of a low salt medium and proper growth regulators
viii) use of molten wax to cover the cut surface

The following experiment was design to investigate the effect of various antioxidants and the use of wax to cover cut surfaces on browning of explants.

Materials:

* banana and coffee shoot cultures; glasshouse grown explant materials of your choice
*100 ml of each of the following antibrowning agents (1 mg/l concentration). Filter sterilised: cysteine, ascorbic acid, PVP and citric acid
*Plain MS medium as Control and MS medium containing the various antibrowning agents, respectively (1 mg/l)
* Sterile petri dishes
* Dissecting instruments

Procedure

1. Nodal cuttings and leaf explants from both *in vitro* materials and glass house material (after surface sterilisation) were taken and cultured onto the various type of medium prepared.

2. Step 1 was repeated with dissection carried out in a sterile petri dish containing the respective antibrowning solution prepared (filter sterilised) earlier and then cultured onto the various type of media. This was to prevent the cut surface coming into contact with oxygen in air during dissection.

3. In both Step 1 and 2, a section of the explant material was cultured immediately onto the respective medium, while another section was left to soak in sterile antibrowning solutions for 30 min.

4. Incubate cultures at 25°C, 16h photoperiod and make observations by scoring intensity of browning in cultures at weekly basis.

Exercise 11: Immobilisation

4.11.1. Encapsulation of somatic embryos

Somatic embryogenesis was first observed in 1958 but the concept of creating artificial seeds using mature embryoids was not published until approximately 20 years later. Initial seed making involved the sandwiching of carrot embryos between wafer disks but recent developments include encapsulation of somatic embryos and dissociated somatic embryos in porous jellies and alginate. Other delivery systems such as fluid drilling and dissociated, non-coated embryos have also been addressed. As the techniques have gained popularity an increasing number of plant species have been used in seed-making e.g. sandalwood, celery, barley. However, the potential value of synthetic seeds for most crops depends on cost structure influenced by the germination potential of embryos, efficiency of plant production and the survival rate of synthetic seeds. Synthetic seeds are more expensive than normal seeds and there must be a value-added component (eg. disease resistance, clonal uniformity, improved nutritional status) to justify the increased expense; only crops for which true seed is currently unavailable are most likely to benefit from artificial seed production, even if at a high cost.

Other crops will need a propagation method faster than conventional seed methods before artificial seeds become competitive.

Problems associated with capsules/artificial seeds

1) The alginate coating is "leaky" in that water-soluble nutrients leach out rapidly.

2) Root and shoot emergence can be hindered.

3) Synthetic seeds can only be stored for a short time as it appears that embryo respiration may be inhibited inside the alginate capsule.

4) Capsules are sticky and cannot be handled efficiently by automatic production and planting equipment. This problem can be solved by coating capsules with a hydrophobic membrane.

5) Capsules dry out rapidly when exposed to air unless coated with an impermeable barrier (e.g. Elvax 4260).

6) Soil conversion rates for synthetic seeds are low (alfalfa - 20%, celery - 10%).

7) Somaclonal variation is a major obstacle for crops where plant-to-plant uniformity is important.

Advantages of synthetic seeds/encapsulation

1) Cells or embryos, packaged in alginate, can be easily transferred to fresh medium, growth regulators can be changed or selected toxins can be applied to the cells. The culture conditions can be changed at will. Beads can be placed into columns or in flat beds and media passed through or over the fixed cells to study the production of secondary compounds by cultured cells.

2) Immobilisation in alginate is useful in the culture of isolated protoplasts which need sequential culture under a range of conditions (eg. gradual reduction of osmoticum).

3) Artificial seeds allow direct planting of propagules into the greenhouse or field, thus by passing acclimatization steps normally associated with micro-propagated plants.

4) When automated systems are developed the combination of high volume production with low cost propagation will be a great advantage.

Hydrogel encapsulation

In order to produce a synthetic seed, the seed coat must be non-damaging to the somatic embryos. It needs to protect the embryos while allowing for germination and conversion (germination). The seed coat must be durable enough to allow ordinary handling during storage, transportation and planting. The coat will need to contain and deliver nutrients, developmental control agents, and other components necessary for germination and conversion. Seed coats could also be supplemented with growth promoting micro-organisms or agricultural chemicals. The capability of the seed coat to allow for monoembryonic seed development is also very important. Finally, the synthetic seed coat must be sowable using existing greenhouse and farm machinery.

Many encapsulation techniques have been attempted but only the gelation method has permitted embryo survival. The most successful gels used are alginate, alginate with gelatin and carrageenin with locust bean gum. In some research the *in vitro* conversion frequencies for synthetic has been equal to those obtained for nonencapsulated embryos. The use of synthetic endosperm did not significantly increase the conversion rate of embryoids.

Materials

* Suspension culture of globular stage carrot somatic embryos
* 200ml MS basal medium + 3% sucrose (**Medium 1**) for culture of encapsulated embryoids
* 200ml solidification solution (**Medium 2**) contained in 50ml flasks

> *Medium 2*:
> MS salts + 80mM $CaCl_2.2H_2O$
> + 3% sucrose

* 200ml culture medium (**Medium 3**):

> ***Medium 3:***
> MS salts + 3% sucrose
> + 30 g/l ("normal viscosity") or 10 g/l ("high viscosity") sodium alginate (polymers of D-mannuronic and L-glucuronic acids).

* Sterile centrifuge tubes
* Sterile petri-dishes
* Sterile plastic Pasteur pipettes or wide bore sterile pipettes.
* Sterile stainless steel filter
* 1 x 100 µm stainless filter (sterile)

Procedures

1. Using the embryoid suspension culture, filter through a 100µm sieve to separate small aggregates from large aggregates. Resuspend the large aggregates in small amount of Medium 1. Only large aggregates are to be used in this experiment.

2. Centrifuge the clusters down (if necessary) in 8ml aliquots placed in centrifuge tubes (650rpm for 5 min).

3. Pour off supernatant and resuspend in Medium 3 at the same level as was previously present in tubes.

4. Fill a sterile plastic pipette with the resuspended alginate/cell suspension mixture and drip (preferably as individual embryos) into a flask containing Medium 2.

5. As the drops come into contact with and sink into Medium 2, a gel matrix is formed thus entrapping cells or embryoids in beads. Allow beads to set (harden) for 30 min and then wash several times on a sterile stainless steel sieve (or tea strainer) with Medium 1.

6. Place beads into 5cm petri dishes with a shallow layer of Medium 1 and culture at 25°C under light conditions.

4.11.2. Immobilisation of embryoids or cells in agarose

An alternative to alginate, which relies on high divalent cations like Ca^{2+} for polymerisation, is agarose which sets at low temperatures (e.g. Type VII, Sigma which sets at 25°C).

Procedure

1. 1g Type VII agarose dissolved in 100ml Medium 1 by heating on a hot plate fitted with a magnetic stirrer and then autoclaved.

2. After cooling the dissolved agarose medium is maintained at 35°C in a water bath.

3. Make up suspension cultures to a cell density of 2×10^4 cells/ml or of somatic embryos.

4. Mix cell suspensions or individual embryos at a ratio of 1:1 with the agarose and stir while at 35°C.

5. 30 - 50ml cell-embryoid/agarose mixture is added to a 100ml flask containing 40ml pure vegetable oil and a teflon stirrer. Maintain at 35°C otherwise agarose will set prematurely.

6. By adjusting the speed of the stirrer globules of agarose containing cells are formed. Increase the stirrer speed until beads of approximately 1mm diameter are obtained.

7. Allow the mixture to cool down rapidly so that the agarose sets. This is best done by placing the suspension into a freezer for a few min followed by 15 min at 4°C in the dark.

8. Decant excess oil and wash beads several times with fresh Medium 1. Centrifugation of beads at 650rpm for 5 min is used to recover the washed beads.

9. Rususpend washed beads in 8ml Medium 1.

10. Pour 8ml bead suspensions into 5cm petri-dishes, seal with clingfilm and incubate in the dark at 25°C.

References

Datla, S., and Potrykus, I., 1989. Artificial seeds in barley: encapsulation of microspore-derived embryos. *Theor. Appl. Genet.* 77: 820.

Eujui, J., Slade, D., Redenbaugh, K. and Walker, K., 1987. Artificial seeds for plant propagation. *TIBTECH* 5: 335.

Janick, J., Sherry, L.K. and Yong-Hwan, K., 1989. Production of synthetic seed by desiccation and encapsulation. *In Vitro Cell Dev. Biol.* 25: 1167 - 1172.

Levin, R., Gaba, V., Tal, B., Hirsch, S. and De Nola, D., 1988. Automated plant tissue culture for mass propagation. *Biol/Technology* 6: 1037.

Redenbaugh, K., Slade, D., Viss, P. and Euyii, J. 1987. Encapsulation of somatic embryos in synthetic seed coats. *Hort. Science* 22(5): 803.

Section 5: GENERAL INFORMATION

5.1. Plant growth substances and plant growth responses

Plant growth regulators (or plant hormones) play key roles in the control of growth in plants. They occur naturally at very low concentrations ($<10^{-9}$ M) within plant tissue and it is at these concentrations that they have their effect on plant growth. Endogenous plant hormones include IAA (indole-3-acetic acid), zeatin, GA_3 gibberellic acid, ABA (abscisic acid) and the gas ethylene. Different hormones work in contrasting ways and have varying effects on plant growth. In plant tissue culture, the addition of exogenous hormones in the medium will supplement, and add to, the naturally occurring endogenous hormone level. Each plant variety will have its own fine balance of hormones depending on the physiology, growth habit and age of the plant, as well as the time of year. In general, cytokinins are shoot inducing substances and are used in plant tissue culture media to promote shoot proliferation. The actual amount of cytokinin used will vary depending on the plant variety. Some herbaceous plants such as tomato, potato and cauliflower, do not need any additional cytokinin to give good shoot growth. In these cases, the endogenous levels present in the small pieces of tissue at explantation is sufficient to support shoot growth and development.

Most woody plants need the addition of a cytokinin in the nutrient medium to give good shoot proliferation. The amount and type(s) of cytokinin used will again depend on the plant variety. A range between 0.1 - 10 mg/l of cytokinin (c. 5×10^{-7}M - 5×10^{-5}M) can be used. Using 0.5 - 2 mg/l (10^{-6} - 5×10^{-6}M) is often a good starting point.

Synthetic cytokinins based on adenine are widely used and include BAP (6-benzylamino purine) and kinetin, phenylurea derivatives like TZ (thidiazuron). By contrast, cytokinins like 2-iP (6-dimethyallylamino-purine) and zeatin are natural ones. These can be used individually or in combination with synthetic ones to obtain desired responses. Other substances such as adenine sulphate have cytokinin-like activity and can also be used. When selecting the type and level of cytokinin, it is important to bear in mind the growth structure of your explant and the subculturing technique to use.

Buds that grow by elongation and produce many nodal units can be cut up simply at internodes. Other plants produce new growth at the base of the plant, for example most monocotyledonous plants.

The naturally occurring auxin IAA (indole-3-acetic acid) is produced by the young leaves to stimulate new root growth. Synthetic auxin are often included in the nutrient medium to stimulate root growth. Some herbaceous plants such as potato, carnation and cauliflower, do not need any additional auxin to induce root growth.

Again, in these cases the endogenous levels in the small pieces of tissue is sufficient. Other herbaceous and many woody plants need the addition of an auxin to give root growth.

In general, a high cytokinin auxin ratio tends to induce shoot growth and a high auxin:cytokinin ratio tends to induce root growth and/or callus depending on the type and strength of auxin used. The actual amount, as well as the ratio of growth substance will form the basis of the plant response and this will differ in different plant species and in different types of explant used.

The three most common synthetic auxins are IBA (indol-3-butyric acid), NAA (1-naphthalene acetic acid) and 2,4-D (2,4-dichlorophenoxyacetic acid). IBA is the auxin which is most commonly used at 0.1 - 2 mg/l to induce a rooting response. NAA can also be used to obtain rooting, or more commonly to induce callus. 2,4-D is used to induce callus and in particular to induce somatic embryogenesis either directly on embryos, leaves, stems etc or indirectly from callus cells.

5.2. Role of vitamins in plants

Vitamins are organic molecules that are needed, in small amounts, by higher animals and plants. Plants are able to synthesise all of their own vitamins but animals rely on certain micro-organisms to synthesise some of the essential compounds. The vitamins serve nearly the same roles in all forms of life. Plant tissues contain both water soluble and fat soluble vitamins, the majority of which are components of co-enzymes.

Water soluble vitamins (Vitamin B complex)

1. Thiamine (Vitamin B_1).

2. Riboflavin (Vitamin B_2) - flavin adenine dinucleotide (FAD) and flavin mononucleotide (co-factor in rooting).

3. Nicotinic acid (nicotinate) - involved in reductive biosynthesis in plants in form of NADPH & NADH. NADH is used primarily for the generation of ATP. (NAD = nicotinamide adenine dinucleotide).

4. Pyridoxine, pyridoxamine (Vitamin B_6) - active in form of pyridoxal phosphate.

5. Pantothenate - pantothenic acid is an important component of Coenzyme A (A = acetylation) essential molecule in metabolism. It is heat stable and is required in many enzyme catalysed acetylations.

6. Biotin - covalently attached to carboxylase enzymes.

7. Folate (folic acid) - component of tetrahydrofolate.

8. Cobalamin (B_{12}) - component of cobamide coenzymes. Cobalt (Co) is an essential component of this coenzyme which is involved in nitrogen fixation in plants.

9. Vitamin C (ascorbic acid).

Fat soluble vitamins (Vitamins A, D, E and K)
Little is known about the molecular basis of fat soluble vitamin action.

1. Vitamin K - required in animals for normal blood clotting. It participates in the carboxylation of glumate residues to X-carboxyglutamate.

2. Vitamin A - precursor of retinal, the light absorbing group in visual pigments. Deficiency results in night blindness.

3. Vitamin D - the metabolism of calcium (Ca) and phosphorous (P) is regulated by a hormone derived from vitamin D.

4. Vitamin E - deficiency in rats leads to infertility. This vitamin protects unsaturated membrane lipids from oxidation.

5.3. Quantification of *in vitro* plant production at Stages II, III and IV.

Micropropagation efficiency

The equation used to determine the potential number of plantlets that can be obtained in a specified time (e.g. in one year) is as follows:

5.2. Role of vitamins in plants

Vitamins are organic molecules that are needed, in small amounts, by higher animals and plants. Plants are able to synthesise all of their own vitamins but animals rely on certain micro-organisms to synthesise some of the essential compounds. The vitamins serve nearly the same roles in all forms of life. Plant tissues contain both water soluble and fat soluble vitamins, the majority of which are components of co-enzymes.

Water soluble vitamins (Vitamin B complex)

1. Thiamine (Vitamin B_1).

2. Riboflavin (Vitamin B_2) - flavin adenine dinucleotide (FAD) and flavin mononucleotide (co-factor in rooting).

3. Nicotinic acid (nicotinate) - involved in reductive biosynthesis in plants in form of NADPH & NADH. NADH is used primarily for the generation of ATP. (NAD = nicotinamide adenine dinucleotide).

4. Pyridoxine, pyridoxamine (Vitamin B_6) - active in form of pyridoxal phosphate.

5. Pantothenate - pantothenic acid is an important component of Coenzyme A (A = acetylation) essential molecule in metabolism. It is heat stable and is required in many enzyme catalysed acetylations.

6. Biotin - covalently attached to carboxylase enzymes.

7. Folate (folic acid) - component of tetrahydrofolate.

8. Cobalamin (B_{12}) - component of cobamide coenzymes. Cobalt (Co) is an essential component of this coenzyme which is involved in nitrogen fixation in plants.

9. Vitamin C (ascorbic acid).

Fat soluble vitamins (Vitamins A, D, E and K)
Little is known about the molecular basis of fat soluble vitamin action.

1. Vitamin K - required in animals for normal blood clotting. It participates in the carboxylation of glumate residues to X-carboxyglutamate.

2. Vitamin A - precursor of retinal, the light absorbing group in visual pigments. Deficiency results in night blindness.

3. Vitamin D - the metabolism of calcium (Ca) and phosphorous (P) is regulated by a hormone derived from vitamin D.

4. Vitamin E - deficiency in rats leads to infertility. This vitamin protects unsaturated membrane lipids from oxidation.

5.3. Quantification of *in vitro* plant production at Stages II, III and IV.

Micropropagation efficiency

The equation used to determine the potential number of plantlets that can be obtained in a specified time (e.g. in one year) is as follows:

$$y = A^n \times B \times F_1 \times F_2 \times F_3$$

where:

- y = Number of plants produced in one year
- A = Number of shoots produced at each subculture ie. multiplication factor
- B = Number of initial cultures available
- n = Number of subculture passages
- F = Loss factors: F_1 = losses in culture from unusable shoots i.e. if 2/10 Explants produce no shoots then F_1 factor = 8/10 = 0.8
- F_2 = Rooting failures, i.e. if 2/10 explants produce no roots then F_2 factor = 8/10 = 0.8
- F_3 = losses on weaning i.e. if 1/10 rooted explants do not survive then F_3 factor = 9/10 = 0.9

This formula is only valid in situations where the whole crop from one initiation is taken through the system and weaned at one time.

Example: Ten shoot tip explants are cultured *in vitro*. Five survive to produce shoots which multiply at a rate of 4 shoots/explant per month (4 weeks). Of the total number of shoots produced 70% are usable for rooting, 90% form roots and 90% can be successfully weaned. There are 12 subculture passages in one year.

Therefore when : $A = 4$, $n = 12$, $B = 5$, $F_1 = 0.7$, $F_2 = 0.9$, $F_3 = 0.9$

$$y = 4^{12} \times 5 \times 0.7 \times 0.9 \times 0.9$$

ie. $y = 4.8 \times 10^7$ (48 million) plants are therefore could be produced in one year (potentially!!)

Calculation using doubling time

The total number of plants produced may be obtained by scoring the numbers of shoots at the beginning (N_o) and end (N) of several successive subcultures. The natural logarithm (*ln*) of the ratio N/N_o is taken and then added to the previous total. The sums of these logarithms (*ln* N/N_o) can then be plotted against the time (y) that has elapsed since the start and the end of the subculture period. In those cases where the plot is a straight line, the increase in shoot number has been exponential, and therefore;

From the graph:

$$K = \frac{\ln \frac{N}{N_o}}{t}$$

$$\ln \frac{N}{N_o} = Kt$$

$$\frac{N}{N_o} = e^{Kt}$$

$$N = N_o e^{Kt}$$

alternatively;

$$t = \frac{\ln \frac{N}{N_o}}{K}$$

Where: K = Slope of the graph

N = number of plants produced at a particular subculture

N_o = number of shoots at the begining

> **Example 1:**
> Predict the shoot doubling time (td) and calculate how many days are required to produce 190,000 plants. Also estimate how many initial explants would be needed to produce 1,000,000 microplants within 9 months. Draw a graph showing the multiplication rate. Given that;
>
> | | |
> |---|---|
> | Number of initial explants (N_o) | = 5 |
> | Number of shoots after 22 days (N_{22}) | = 15 |
> | Number of shoots after 44 days (N_{44}) | = 45 |
> | Number of shoots after 66 days (N_{66}) | = 100 |
> | Number of shoots after 88 days (N_{88}) | = 206 |

Thus,
After 22 days:

$$\ln\frac{N_{22}}{N_o} = \ln\frac{15}{5} = 1.099$$

After 44 days:

$$\ln\frac{N_{44}}{N_{22}} = \ln\frac{45}{15} = 1.099$$

After 66 days:

$$\ln\frac{N_{66}}{N_{44}} = \ln\frac{100}{45} = 0.799$$

After 88 days:

$$\ln\frac{N_{88}}{N_{66}} = \ln\frac{206}{100} = 0.723$$

A graph of the **sum of $ln(N_o/N)$** (y-axis) can then be plotted against the time (x-axis) that has elapsed since the start of the experiment.

Then, from the graph:

$$K = \frac{3.720 - 2.198}{4 - 2} = 0.761$$

If N = 190,000 and N_o = 5, then;
ie. if each subculture period were 22 days in length, then 308 days would be required to produce 190,000 plants from the 5 original cultures at the levels of multiplication measured.

$$190,000 = 5e^{Kt}$$
$$e^{Kt} = 38,000$$
$$\ln e^{Kt} = \ln 38,000$$
$$Kt = \ln 38,000$$
$$t = \frac{\ln 38,000}{K} = \frac{\ln 38,000}{0.761}$$
$$t = 13.9$$

Shoot doubling time (td), ie. in the case of micropropagation; the time it takes for a given amount of shoot material to double its bud potential. It can be obtained from the equation;

$$N_{td} = 2N_o$$

Whereas,

$$t = \frac{\ln \dfrac{N}{N_o}}{K} = \frac{\ln \dfrac{2N_o}{N_o}}{K}$$
$$t_d = \frac{\ln 2}{K} = \frac{0.6931}{K}$$

Therefore, for this experiment, the value of 'K' from the graph is 0.761. Thus;

$$t_d = \frac{0.6931}{0.761} = 0.911$$

which gives a shoot doubling time $(0.911 \times 22) = 18$ days; a fraction of one subculture period of 22 days.

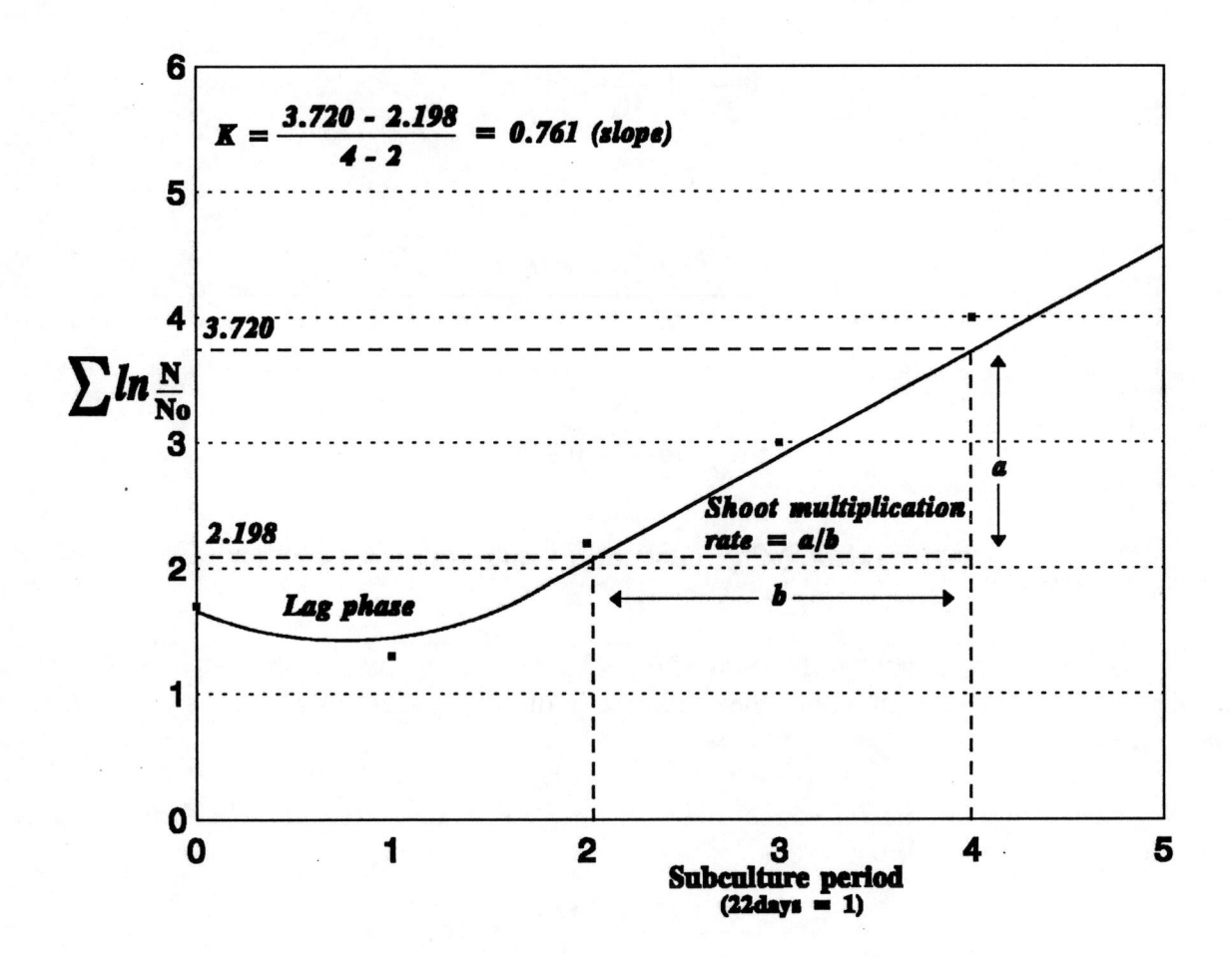

Figure 4. Graph showing the growth/multiplication rate of micropropagated plants

> **Example 2:**
> Given the number of initial explants (N_o) equal to 10, and the number of shoots after one month (N_1) equal to 20. Calculate the value of 'K' and the shoot doubling time (td) without plotting a graph.

Since: $N_o = 10$
$N_1 = 20$
$N_2 = 50$
$N_3 = 125$

Thus;

after 1 month:

$$\ln\frac{N_1}{N_o} = \ln\frac{20}{10} = 0.693$$

after 2 month:

$$\ln\frac{N_2}{N_1} = \ln\frac{50}{20} = 0.916$$

after 3 month:¾

$$\ln\frac{N_3}{N_2} = \ln\frac{125}{50} = 0.916$$

Therefore the **sum of *ln(N/No)*** after the 2nd subculture is 1.609 and after the 3rd subculture is 2.525.

If we imagine a graph of the **sum of *ln(N/N$_o$)*** plotted against the time; then the value of K (the slope) at any point will be the difference in the **sum of *ln(N/N$_o$)*** over the difference in the time lapse.

Therefore, the gradient (slope) between the 3rd subculture and the 1st subculture (2 month of time lapse) will be;

$$K = \frac{2.525 - 0.693}{3 - 1} = \frac{1.832}{2} = 0.916$$

And therefore, the shoot doubling time (td) for this experiment is 23 days (*ca.* ¾ of a month).

$$t_d = \frac{\ln 2}{K} = \frac{0.6931}{0.916} = 0.76$$
$$t_d = 0.76 \times 30 = 23 \, days$$

Analysis of plant tissue growth rates

A wide choice of parameters for the analysis of growth rates of various organs are available.

Roots:
* Increase in length of cultured roots. Cultured roots do not normally exhibit secondary thickening and the cross sectorial area remains constant during growth.
* Increments in length represent an accurate measure of volume change with growth.

Callus:
* Increases in fresh weight are frequently used.
* Measurement of the dry weight of callus can be used as an indication of the cultures biosynthetic activity.
* Dry weights of cell suspensions can be determined. Measurement of the total amount of cell mass in a given volume of culture can be used for suspension cultures. The cell mass is obtained by centrifugation (Packed Cell Volume). Another indication of growth is the increase in cell number of the culture (cell density).
* The mitotic index (MI) represents the percentage of the total cell population of a culture that exhibits the same stage of mitosis.
* Tracheary element counts in primary explants after maceration of the primary explants.

5.4. Statistics

Data obtained from plant tissue culture research may often be difficult to analyse statistically. The data is often subjective and qualitative rather than quantitative. Statistical analyses are on the other hand an essential part of any biological research. With the availability of a large number of statistical methods for biological researchers, one should avoid complicated statistical procedures. Remember, statistics should be used as a tool to compare treatments of interest and should not dictate the treatments.

Experimental designs should take into account the eventual analysis, otherwise it may not answer the desired questions when analysed. Therefore, do spent some time before conducting an experiment to plan an experimental design and analysis.

Induction of shoot/root formation from embryogenic callus can often only be analysed as yes/no response. Often the percentage is very low and also not necessarily following a normal distribution. On the other hand you may get a variety of responses including numerous small shoots that are very difficult to count.

> **In such cases you can employ a scoring system**
> 1............................. 5
> no. shoots..................many shoots
> *(with subjective estimate)*

Often the lack of material is a severe handicap, especially if you do not have much material because the plant is rare or difficult to propagate.

If only a small amount of material is available, replications will be low. If there is sufficient material for a factorial experiment it is best to vary only one or two factors at the most.

2 factor experiment.

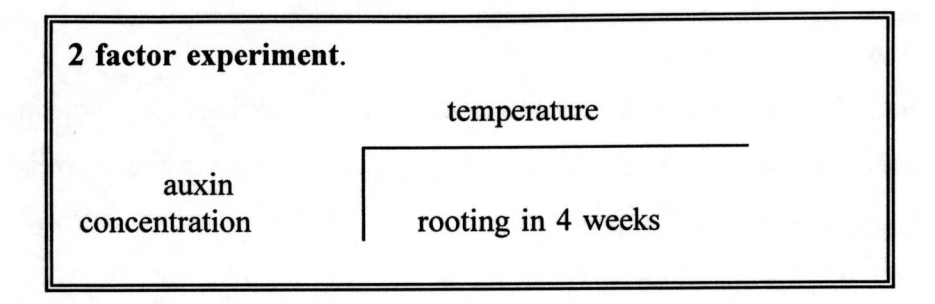

	temperature
auxin concentration	rooting in 4 weeks

This data could be analysed appropriately using ANOVA, REGRESSION etc.

Experimental design:

Tubes are useful as one plant/tube can increase replication and makes use of less medium. Randomise or block the treatments in the polystyrene trays. A computer package can also be used to carry out ANOVA and REGRESSIONS and to draw simple block graphs.

Quantitative parameters

Shoot Production
  * Multiplication rates - no. of propagules/shoot/week
  * Shoot doubling times - no. of days to double propagule number

Root Production
  * Percentages - % rooting after X weeks
  * only applicable when only a few roots(1- 10) are produced.

> ***Fresh weights***
> * shoots and callus can be weighed under sterile conditions on the electronic balance that can be swabbed down and placed in a sterile cabinet. Weights and growth rates can be numerically compared and analysed statistically.
> * roots have to be removed for weighing, so only useful if the rooted shoot is not required for weaning

> ***Lengths***
> * shoot lengths can be measured aseptically in the flow cabinet by placing graph paper under the sterilised polypropylene.
> * length of longest roots for a shoot with many roots eg. carnation.

Qualitative parameters

- general description
- photography - a visual record
- histology - an investigation into the morphogenesis of the plant material e.g. to assess organogenesis or embryogenesis

Reference on suitable statistical methods for plant tissue culture work

1. Carl W. Mize & Young Woo Chun (1988). Analysing treatment means in plant tissue culture research. *Plant Cell, Tissue and Organ Culture* **13**: 201-217.

2. Michael E. Compton (1994). Statistical methods suitable for analysis of plant tissue culture data. *Plant Cell, Tissue and Organ Culture* **37**: 217-242.

5.5. Laboratory layout, equipment and costing

(A) Facilities

* ***Medium preparation/washing up area.***

    This is a clean area free from excessive dirt and air currents where media are prepared and glassware is cleaned and stored. Smooth floor preferably lined with linoleum or other suitable washable surface.

* ***Sterilisation equipment***

    Storage space for chemicals, glassware and other hardware.

* ***Transfer area***

    Clean, dust-free area where initiation and subculture of plant material can be carried out.

* ***Culture incubation area***

    Space for incubation of cultures on shelves illuminated with appropriate quality and intensity of light.

    Stages I, II - ca 600 - 3,000 lux (ca 1-3 W m^{-2})

    Stages III, IV - 2,000 - 10,000 lux (ca 12-25 W m^{-2})

    Ambient temperatures (25°C) inside area must be held at controlled levels preferably within ± 2°C.

* ***Equipment for the examination and evaluation of the cultures and also good record keeping facilities are essential***

* ***Plant establishment area***

    A protected nursery area where plantlets can be hardened off before being transferred to unprotected environments. Fogging systems operated in conjunction with plastic covered benches are the most usual types of setup employed.

(B) Layout of micropropagation laboratory

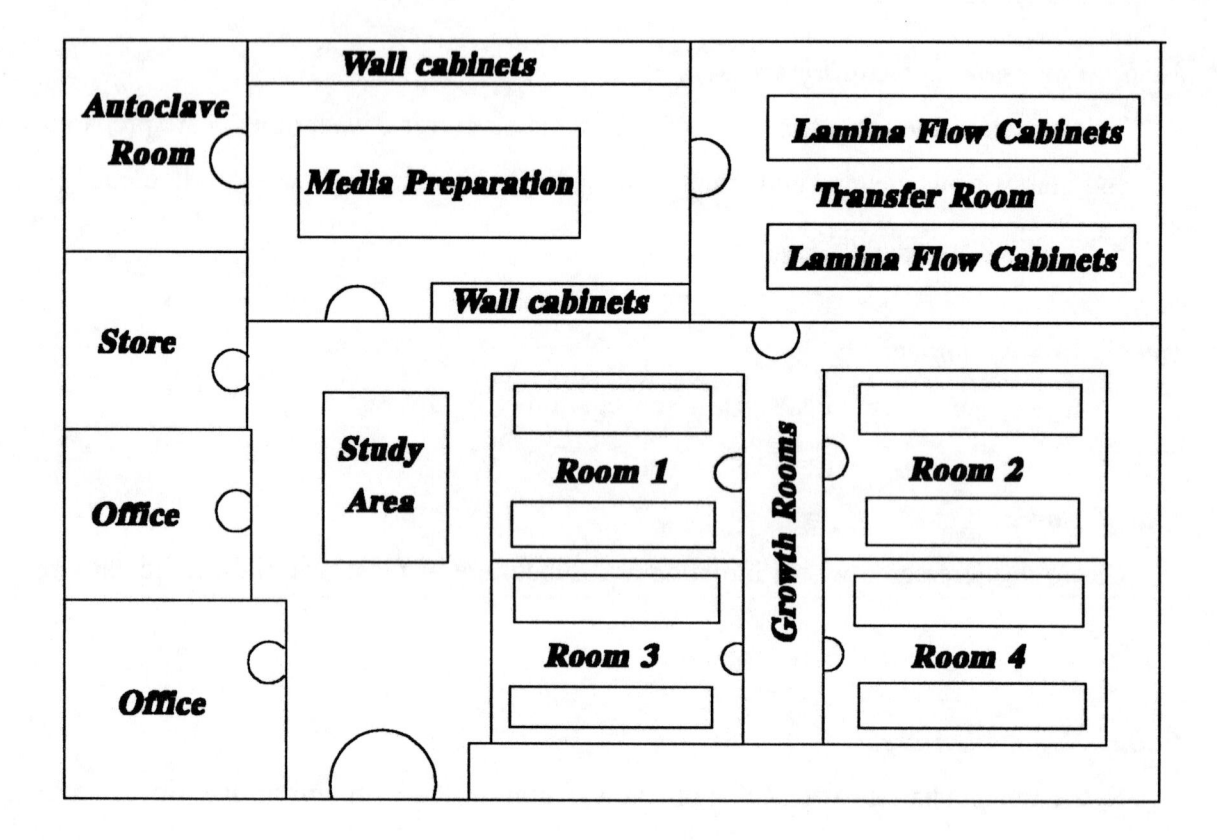

Figure 5. Floor plan of a typical *in vitro* propagation laboratory with the necessary growth rooms. It is designed in such a way to be self- contained with as little opening to the outside as possible. The normal flow of materials and operations for *in vitro* procedures are taken into account during the designing of this laboratory.

(C) Equipment and materials

* *Medium preparation area requires*:

Glassware: (beakers, flasks, measuring cylinders, petri dishes culture jars, tubes of various sizes); Pyrex and borosilicate preferable

Chemicals: (prepared powdered media or individual salts, alcohol for swabbing down) pH meter (or pH papers); Analar grade

Refrigerator: (for storage of some stock solutions and chemicals susceptible to breakdown in storage at normal room temperatures); domestic types generally adequate

Balances: (to weigh out accurate amounts of media constituents. Top pan balances preferable to bar types), eg. Sartorius.

Distillation: (demineralisation) apparatus for producing adequate amounts of clean, purified water, eg. Millipore

Equipment for **filter sterilisation** of heat sensitive growth regulators and other compounds

Covers: Cotton plugs/aluminium foil/metal lids for covering culture jars during sterilisation. Hot plate (domestic type to facilitate dissolving salts and agar and for autoclaving small amounts of media in pressure cookers)

Autoclaves: (domestic pressure cookers or specialised autoclaves). Magnet/heater stirrers (to facilitate mixing) during heating of agar

* *Culture transfer area*
    - Laminar flow cabinets (culture cabinets)
    - Dry heat sterilisers (for sterilising instruments during culture transfer work). These can be bunsen burners, meths burners or steribead sterilisers
    - Dissection instruments (scalpels, forceps, scissors)
    - Glass plates (for cutting surfaces)
    - Alcohol (for swabbing down work surfaces)
    - Labelling pens (or labelling machines in large operations)
    - Trays/trolleys
    - Polypropylene covers/filter papers (previously sterilised by autoclaving)

* *Cleaning and washing facilities*
    Glassware washed in hot detergent bath and then repeatedly rinsed in tap water followed by distilled water. Glassware oven dried before re-use

* *Culture incubation area*
    Dexion or angle iron framework supporting white wood/plastic shelving.

    Lights: (fluorescent strips of appropriate types e.g. Phillips 'TL' D83 strip lights).

    Air circulator fans: (to ensure equable temperature throughout room); cooling and refrigerator units. Air conditioner of domestic type may be suitably adapted

    Electrical chokes: circuitry external to area to reduce heating levels

    Timer: (to set photoperiod - usually 16 hours)

* *Plant establishment area*
    - Fogging/mist propagation systems, mobile benching, supplementary lighting (mercury lamps), plastic covers supported by metal hoops

- Suitable compost mixtures or soilless substrates in seed trays or individual pots

* **Additional equipment**:
    - Bench top centrifuge for protoplast purification; fire extinguisher; first aid kit

(D) Costs of establishing micropropagation facilities

Three levels considered:
- a) Small exploratory unit in domestic setting
- b) Medium unit in commercial nursery setting
- c) Large unit for specialisation in microplant production

> **Small (exploratory) unit**:
> Assume: Prepackaged media used (no weighing)
>   pH papers
>   Single distilled water unit
>   Still-air cabinet
>   Meths burners or gas burner
>   Basic dissection equipment
>   24 sq.ft. dexion light unit
>   Honey jars
>   A pressure cooker/domestic kitchen stove

Approximate total cost £800

This unit could produce ca 2,000 - 5,000 plantlets per month of an amenable nursery stock. Recurrent expenditure of ca 25 - 60 per month.

> **Medium sized unit**:
>
> Assume: Top-pan balance
>
> Refrigerator
>
> pH meter (£200)
>
> Honey jars
>
> Basic dissection instruments
>
> Portable gas burner
>
> 2 pressure cookers
>
> A 1 metre laminar airflow cabinet (£1500)
>
> Single distilled water still
>
> 30 m^2 culture area

Approximate total cost £4,800

This unit could produce 10,000 - 15,000 plantlets per month of amenable types of nursery stock. Recurrent expenditure (excluding labour) ca £150 per month.

> **Large scale unit**:
>
> As in b) but with: 3 x 3m laminar flow cabinets (£4000)
>
> 3 x gas burners or "Steribead" sterilisation units
>
> A double distilled water supply (4 litres per hour)
>
> Ample glassware/honey jars
>
> Disposable culture ware
>
> Air-conditioner unit for culture area (£5800)
>
> Large autoclave (£9,500)

Approximate total cost £22,500

This unit would be capable of producing 40,000 - 50,000 plantlets of amenable types of nursery stock per month. Recurrent expenditure (excluding labour) *ca.* £450 per month.

5.6 Laboratory equipment suppliers 1994

Astell Scientific
Powerscroft Road,
Sidcup,
Kent. DA14 5EF.

Autoclaves

British Sterilizer Co. Ltd.
15 Roebuck Road,
Hainault,
Ilford,
Essex. IG6 3TX.

Autoclaves

Priorclave Ltd.,
129 Nathan Way,
Woolwich,
London, SE28 0AB.

Autoclaves

Bassaire Ltd.
Duncan Road,
Swanick,
Southampton. SO3 7ZS.

Air flow cabinets

BDH Chemicals
Freshwater Road
Dagenham
Essex. RM8 1RZ.

Laboratory Chemicals:-
Macro & micro elements
Tween 20
Buffer Tablets

Camlab Ltd.
Nuffield Road
Cambridge. CB4 1TH.

General Laboratory
Syringes

Cannings Parry Packaging
Avonmouth Way
Avonmouth
Bristol. BS11 9DZ.

Packaging materials
Polypropylene sheets

Denley Instruments Ltd.
Natts Lane
Billingshurst,
Sussex. RH14 9EY.

General laboratory
Aluminium trays

Downs Surgical Ltd.
Church Path
Mitcham
Surrey. CR4 3EU.

Surgical suppliers
Instruments, rubber gloves
scalpel blades

Flow Laboratories
Harefield Road
Woodcock Hill
Rickmansworth
Hertfordshire WD3 1PQ.

Tissue culture suppliers
7x detergent
MS media
Airflow cabinets

Gallenkamp Ltd.
Southern Region Sales Office
Belton Road West
Loughborough
Leicestershire LE11 OTR

General laboratory supplies
pH meter, hot plate,
stirrers, pressure cookers,
drying oven, autoclave
tape, aspirators,
instrument oven, spatulas, pi-pumps,
alochol bottles, glass rods,
weighing boats, filter papers,
wash bottles, thermologs,
thermometers, reverse osmosis
unit, trolleys, alcohol basket,
safety goggles and masks, timer,
bottle brushes, magnetic
followers, beakers, flasks,
measuring cylinders, Schott bottles,
pipettes and all glass and
plasticware

Griffith and Neilson Plastics
Wyvern House,
Station Road
Billingshurst,
Sussex. RH14 9SE.

Plastic disposables

Arnold R. Horwell
73 Maygrove Road
West Hampstead
London NW6 2BP

General lab/medical
Forceps especially

Imperial Laboratories
West Portway,
Andover,
Hants. SP10 3LF.

Cell culture products
Glass bead steriliser
MS media

Jencons (Scientific) Ltd.
Cherrycourt Way Industrial
Estate, Stanbridge Road,
Leighton Buzzard,
Bedfordshire. LU7 8UA.

General laboratory supplies
Nescofilm, shakers

Jouan Ltd.
130 Western Road, Tring, Herts. HP23 4BU.
Harry Hodds,
Sorbarod systems,
Friary Chambers,
Whitefriargate, Hull. Tel: 0482 23548

Media dispensers

Millipore (UK) Ltd.
Millipore House,
11 - 15 Peterborough Road,
Harrow, Middx. HA1 2YH.

Filters

Oxoid Ltd.
Wade Road
Basingstoke
Hampshire. RG24 OPW.

Microbial & Diagnostic
Reagents, Agar, Al caps
for tubes

Richardsons of Leicester
Evington Valley Road
Leicester, Leics. LE5 5LJ.

Tissue culture supplies
Jars, tubes, containers

Sartorius
Longmead
Blenheim Road,
Epsom, Surrey. KT19 9QN.

Balances

Seward Medical
131 Great Suffolk Street
London, SE1 1PP.

Surgical equipment
Instruments, rubber
gloves, scalpel blades

Whatman Labsales Ltd.
Coldred Road,
Parkwood
Maidstone, Kent. ME15 2BR.

General laboratory
supplies,
Filter paper

Carl Zeiss (Oberkochen) Ltd.
P.O. Box 78,
Woodfield Road,
Welwyn Garden City, Herts. AL7 1LU.

Microscopes

For Growth Rooms, local firms may be helpful in constructing racks, ballasts, etc. Lights are Philips TLD 50W/84HF.

> **Refrigeration firms include:**

Hussmann Refrigeration Ltd.,
Unit 7, Quarry Wood Industrial
Estate, London Road, Maidstone,
Kent. ME20 7NA.

Toomeys Refrigeration
Tyland Corner,
Sandling,
Maidstone, Kent. ME14 3BH.

Newey & Eyre   Lamps and ballasts
Unit 1, Cotton Road,
Wincheap Industrial Estate,
Canterbury, Kent. CT1 3RB.

> **Equipment required which should be available locally, or can be made:**

Marker pens
Handcream
Hot ring
Sieves
Aluminium foil
Clingfilm
Labels
Mops
Chairs
Stools
Label gun
Scissors
Masking tape
Greaseproof paper
Paper bags
Polythene bags
Paper towelling (rolls)
Holder for above
Tape dispensers
Refrigerator (also Jencons)
Aluminium boxes and trays
Rubber bands
First Aid Kit
Honey jars, Sugar, Alcohol - refer to local Customs & Excise

Bowls
Savlon
Glass plates
Spray bottles
Laboratory coats
Oven gloves
Bins
Bleach
Towels
Tea Towels
Dettox
Buckets

Acknowledgements

Thanks to everyone who either directly or indirectly contributed to this revised edition of the UAPS MANUAL. We are especially grateful to Bridget Cue for typing the original manuscript of the manual.

Sinclair Mantell, Reader in Horticulture

Kodi Kandasamy, Short Course Tutor
Plant Tissue Culture and its Uses in Micropropagation and Plant Biotechnology

14:1:95